Mind the Body

Our own body seems to be the object that we know the best for we constantly receive a flow of internal information about it. Yet bodily awareness has attracted little attention in the literature.

Mind the Body is the first comprehensive treatment of bodily awareness. Frédérique de Vignemont seeks to answer questions such as: What makes the awareness of our own body so special? How do I perceive it and what makes me feel this specific body as my own? Can I fail to feel that it belongs to me? Or conversely, can I take someone else's body for my own? Can I extend my body beyond my biological boundaries and incorporate a tool, a prosthesis or an avatar's body, for instance? To answer these questions, we need a better understanding of the various aspects of bodily self-awareness, including the spatiality of bodily sensations, their multimodality, their role in social cognition, their relation to action, and to self-defence.

This volume combines philosophical analysis with recent experimental results from cognitive science, leading us to question some of our most basic intuitions.

Frédérique de Vignemont is a CNRS senior researcher and the deputy director of the Jean Nicod Institute in Paris. Her research is at the intersection of philosophy of mind and cognitive science. Her major current works focus on bodily awareness, self-consciousness, and social cognition. She has published widely in philosophy and psychology journals on the first-person, body schema, agency, empathy, and more recently on pain. She has edited interdisciplinary volumes on the body and the self, *The Subject's Matter* (MIT Press, 2017) and on peripersonal space, *The World at Our Fingertips* (OUP, forthcoming). She is the recipient of the 2015 Young Mind & Brain prize. She is also one of the executive editors of the *Review of Philosophy and Psychology*.

Mind the Body

An Exploration of Bodily Self-Awareness

Frédérique de Vignemont

OXFORD
UNIVERSITY PRESS

OXFORD
UNIVERSITY PRESS

Great Clarendon Street, Oxford, OX2 6DP,
United Kingdom

Oxford University Press is a department of the University of Oxford.
It furthers the University's objective of excellence in research, scholarship,
and education by publishing worldwide. Oxford is a registered trade mark of
Oxford University Press in the UK and in certain other countries

First Edition published in 2018
First published in paperback 2020

Published in the United States of America by Oxford University Press
198 Madison Avenue, New York, NY 10016, United States of America

British Library Cataloguing in Publication Data
Data available

Library of Congress Cataloging in Publication Data
Data available

ISBN 978-0-19-873588-5 (Hbk.)
ISBN 978-0-19-885802-7 (Pbk.)

Contents

Part II. Body-Builder

Part III. Bodyguard

Acknowledgements

I feel few people have been as lucky as I have been in my philosophical career. Through the years I have had the chance to meet incredibly sharp and creative philosophers and psychologists who have taught me all that I know. Not only were they wonderful teachers but they have also become precious friends. This manuscript is the outcome of our discussions. I am extremely grateful to them all. In particular, I would like to thank those without whom I would not be where I am, especially Ned Block, Denis Forest, Uta Frith, Alvin Goldman, Patrick Haggard, and Pierre Jacob. I would also like to thank those with whom I have spent hours arguing, debating, and laughing, especially Adrian Alsmith, Tim Bayne, Jérôme Dokic, Alessandro Farnè, Olivier Massin, Myrto Mylopoulos, Elisabeth Pacherie, and Hong Yu Wong. There are also all those whose work has inspired me and whose feedback has played a key role in this book, especially Yochai Ataria, Malika Auvray, Elena Azanon, David Bain, José Luis Bermúdez, Anna Berti, Alexandre Billon, Olaf Blanke, Lucilla Cardinali, Marcello Constantini, Abraham Sapien Cordoba, Henrik Ehrsson, Shaun Gallagher, Francesca Garbarini, Adrienne Gouzien, Marie Guillot, Nick Holmes, Giandomenico Iannetti, Marc Jeannerod, Marjolein Kammers, Colin Klein, Uriah Kriegel, Jérémie Lafraire, Tim Lane, Matt Longo, Christophe Lopez, Asifa Majid, Tamar Makin, Alisa Mandrigin, Tony Marcel, Mike Martin, Thomas Metzinger, Alejandro Mujillo, Matthew Nudds, Kevin O'Regan, Chris Peacocke, Victor Pitron, Daniele Romano, David Rosenthal, Andrea Serino, Carlota Serrahima, Tania Singer, and Manos Tsakiris. Finally, there are those who have been so important to me along the way that they do not need to be named.

List of Figures

List of Tables

Introduction

Now there is every reason for expecting it to be a *highly unusual relation*. For we are talking about the epistemological link between a man and in all probability himself qua material object! and in any case the one object in the universe that is his own body. Is not that likely to be unusual?

(O'Shaughnessy, 1980, vol. 1, p. 138)

The relationship between the body and the self raises a number of major issues in philosophy. First, one may ask what the body is for the self. Is the self embodied or purely mental? Does the body guarantee personal identity? One may also ask which body has this relation to the self. One assumes that there is only one specific body that belongs to the subject but which one and how does one individuate it? One may further inquire about the ethical and legal implications of one's relation to the body. What proprietary rights does one have towards one's body? What is its moral status and does it vary whether it is the living body or the body after death? Some of these questions have been at the core of philosophical debates for centuries, others have become crucial only more recently with the use of human parts for biomedical research. Although they are fascinating, I will leave them aside here and start with the most basic way one encounters the "unusual" relation between the body and the self, namely with the philosophical exploration of *bodily awareness*.

Consider the following basic example: I touch the table with my hand. My tactile sensation includes sensations of resistance, texture, and temperature, as well as the sensation of the location at which I feel the pressure to occur, namely the hand. I am also aware that the hand on which I feel touch is *mine*. This type of self-awareness is known as the sense of bodily ownership, for want of a better name. It might seem indeed that I do not "own" my body; I only own my laptop, my flat, and my books. I have a more privileged relation with my body than with any other objects, one might even say a relation of identity. The fact is that numerous other languages—although not English—use different suffixes to indicate the possession of alienable (e.g. my flat) and inalienable entities (e.g. my hand) (Kemmerer, 2014). But the relation that we have with our body cannot be characterized only by the fact that it is inalienable. Indeed I can also qualify my relation to my son in the same way. We thus need to go beyond in our description and ask what it actually means to experience one's body as one's own. Although introspectively familiar, it is hard to pinpoint exactly the nature of the specific relationship that one has uniquely with the body that one experiences as one's own. Whereas the sense of agency has been systematically explored, taking on board puzzling disorders as well as computational models of action control (Bayne, 2008; Jeannerod and Pacherie, 2004; Synofzik et al., 2008; Roessler and Eilan, 2003), the sense of bodily ownership has been largely neglected in philosophy. Yet these last twenty years have seen an explosion of experimental work on body representations, which should help us shape and refine our theory of bodily awareness.

The Mind in the Body

Nowadays, it is often assumed that if you are interested in the body, you must be a proponent of embodied cognition. Embodied theorists indeed try to elevate the importance of the body in explaining cognitive activities. They study the mind not in isolation, as if it were a disembodied computational device or a brain in a vat, but in its mutual interactions with the body and the world. They claim that the body has a crucial significance in how and what the organism thinks and feels. According to Gallagher (2005) in *How the Body Shapes the Mind*, embodiment affects not only perception, emotion, and action but also higher mental processes:

nothing about human experience remains untouched by human embodiment: from the basic perceptual and emotional processes that are already at work in infancy, to a sophisticated interaction with other people; from the acquisition and creative use of language, to higher cognitive faculties involving judgment and metaphor; from the exercise of free will in intentional action to the creation of cultural artefacts that provide for further human affordances.

(Gallagher, 2005, p. 247)

Given this surge of interest in embodied cognition from philosophers and cognitive scientists, it would seem legitimate that they focus part of their interest on the body itself and that one should be able to find some answers about the way we experience our body in this growing programme of research. But although embodied theorists criticize classical cognitivism for its neglect of the body, one may wonder whether they do a better job in accommodating the body than their rivals. Even the so-called embodied approach takes the body for granted and rarely explicitly investigates it.

The fact that there are only a few descriptions of bodily experiences in embodied theories is understandable if the notion of embodiment—a notion too often left undefined—refers to the physical body in interaction with its environment. And indeed some embodied theorists are primarily interested in what we can do with the body, instead of how we experience it (Chemero, 2009; Gallagher, 2008; Thelen and Smith, 1994; Turvey and Carello, 1995; van Gelder, 1995). They give a primacy to action. Hence, what matters for them is the physical body insofar as it allows a fluent adaptive coupling with the environment. The body is only conceived of as the necessary intermediary between the mind and the world and there is actually no need to mentally represent it (Kinsbourne, 1995, 2002). Modifying an expression coined by Brooks (1991), they may say that the body is "its own best model". As Merleau-Ponty (1945) notes, the body has a feature possessed by no other object in the world: it never leaves us. Why then would we need an internal representation of it when we could simply retrieve the relevant information about it as and when we need it? If biological systems only know what they need to in order to get their job done (Clark, 1989), is the body really something that the brain needs to "know"?

It may be true that classical cognitivism has failed to realize the significance of bodies in cognitive processes. Yet this neglect should not necessarily lead to a radical change of paradigm and complete rejection of the notion of mental representation (Clark, 2008; Shapiro, 2010), especially mental representation of the body. Consider the recent discovery of mirror neurons and the subsequent embodied approach to social cognition (Goldman and Vignemont, 2009; Gallese and Sinigaglia, 2011). It has been shown that the same neural resources are involved both when one executes an action and when one sees another individual performing the same action (Grezes and Decety, 2001). It has been argued that this type of mirror system plays a role in the understanding of other people's actions. I shall come back to these mirror systems in Chapter 7, but it is worth noting here that one understands the other in virtue of *representing* the movement in the motor system, and not in virtue of performing the same movement. Contrary to what some radical embodied proponents may believe, mental representations can be both a more useful and a more parsimonious solution than repetitive inter-actions. This is so if one accepts that representations can be highly dynamic, nonconceptual, and even action-orientated. Cognition can then be said to be embodied because it is affected by the way the body is represented in the mind. But if the thesis of embodied cognition is that bodily representations play an essential role for most of our cognitive abilities, then it is of extreme importance to understand how the body is represented.

To conclude, the body has come back by a side door with embodied cognition, but what has embodied cognition told us about the body? Proponents of embodied cognition often disagree on the degree and the nature of the involvement of the body in the mind. I have proposed here two interpretations of the notion of embodiment, one of which appeals to physical bodily activities and the other to mental representations of the body. To limit oneself to the former would reduce the scope of the embodied approach, missing the potentially important role of bodily representation for cognition. This view may displease many embodied theorists (for example, de Bruin and Gallagher, 2012). They may consider that positing body representations actually undermines the explanatory role of the body. One consequence of this view is indeed that the cognitive abilities of a brain-in-a-vat could be embodied: one would end up with some mental representations (of the body) affecting other mental repre-sentations in the closed-loop of the mind. The question is whether this is a real problem. I doubt it. But it means that in order to know what role bodily representations can play in our mental life, one first needs to investigate how we represent our body. The embodied approach claims to return the mind to the body. Here, I shall return the body to the mind.

The Body in the Mind

Let us leave aside what the body can or cannot do for cognition and focus on what it feels like to have a body. Our own body may seem to be the object that we know the best for we constantly receive a flow of information about it through what I call bodily senses. These include five informational channels that provide direct internal access to our own body. *Touch* is mediated by cutaneous mechanoreceptors. It carries information about the external world (e.g. shape of the touched object), but also about the body (e.g. pressure on the specific part of the skin) (Katz, 1925). *Proprioception* provides information about the position and movement of the body. It includes muscle spindles, which are sensitive to muscle stretch, Golgi tendon organs, which are sensitive to tendon tension, and joint receptors, which are sensitive to joint position. *Nociception* responds to dangerously intense mechanical, mechanothermal, thermal, and chemical stimuli. According to the dominant theory, noxious signals are inhibited, enhanced, or distorted by various factors via a gating mechanism at the level of the spinal cord that controls the signals from the periphery to brain structures and via a central gating mechanism (Melzack and Wall, 1983). *Interoception* provides information about the physiological condition of the body in order to maintain optimal homeostasis (Sherrington, 1906). Interoceptive signals arise within four systems: cardiovascular, respiratory, gastrointestinal, and urogenital. The *vestibular system* in the inner ear provides information about the balance of the body. It includes three roughly orthogonal semicircular canals, which are sensitive to motion acceleration as our head moves in space, and two otolith organs, which are sensitive to the pull of gravity.

Yet, despite all these sources of information, the phenomenology of bodily awareness seems surprisingly limited. It appears as less luxuriant and detailed than the phenomenology of visual awareness, which can be analysed as full of fine-grained colour shades and well-individuated 3D shapes that move around. It seems at first sight reducible to the "feeling of the same old body always there" or to a mere "feeling of warmth and intimacy" (James, 1890, p. 242). For instance, while typing on a laptop, we do not vividly experience our fingers on the keyboard and our body seems only to be at the background of our awareness. Our conscious field is primarily occupied by the content of what we are typing, and more generally, by the external world rather than by the bodily medium that allows us to perceive it and to move through it (Gurwitsch, 1985). We use the body, but we rarely reflect upon it.

Nonetheless, when our body becomes less familiar, we can grasp the many ways our body can appear to us. One might believe that we could not be wrong about our body because we have a privileged access to it through bodily senses that put us in direct contact with it, but there are so many bodily illusions and it is almost impossible to provide an exhaustive list of them (see Appendix 1 for a partial attempt). It can take no more than going to the dentist to experience one of them: after dental surgery, your mouth often feels bigger, although it looks normal. This is a surprising side effect of anaesthesia (Gandevia and Phegan, 1999; Türker et al. 2005). The Pinocchio illusion is also quite striking (Lackner, 1988). If the tendons of your arm muscles are vibrated at a certain frequency, you experience illusory arm movements. You feel, for instance, your arm moving away from you if your biceps tendon is vibrated, and if you simultaneously grasp your nose, you experience your nose as elongating by as much as 30 cm. The Pinocchio illusion results from a sensorimotor conflict between erroneous proprioceptive information (i.e. your arm moving away from you) and accurate tactile information (i.e. contact of your nose and your fingers). But it may be the experience of phantom limbs in amputees that best brings the phenomenology of bodily awareness into the limelight. Whereas our body generally stays at the margin of consciousness, phantom limbs are at its foreground. A patient, for example, reported: "I, says one man, I should say, I am more sure of the leg which ain't than of the one that are, I guess I should be about correct" (Mitchell, 1871, p. 566). By analysing phantom limbs, we may thus shed light on what it is like to have a "real" body.

Many amputees experience from the inside the continuous presence of their lost limbs. They feel their shape and size, they feel them at specific locations and in specific postures, they feel sensations in them, sensations of cold and heat, for instance, and too often sensations of pain, and in some cases, they feel them moving and directly obeying their will. Finally, they feel that the phantom limb is part of their body. These various aspects of bodily awareness are familiar to us, even though not always vivid in our consciousness. What are lesser-known are the following puzzling results, which raise fundamental questions about bodily awareness.

Amelic patients and supernumerary phantom limbs: Some individuals are born with missing limb(s) and yet they can experience realistic phantom limbs, although they have never moved them or seen them

(Melzack, 1990). By contrast, other patients experience the presence of phantom limbs although they do not miss any limb and they can feel, for instance, as if they had three arms and three legs (Staub et al., 2006).

How do we build up the representation of our body? We constantly receive a flow of information through external perception and through internal perception, the latter being active all the time. However, if amelic patients can experience phantom limbs, there must be a further basis upon which the representations of the body are built. Is there a kind of innate template of the human anatomy (e.g. two arms, two legs)? But how is such an anatomical template compatible with the experience of supernumerary phantom limbs? Is there no limit to the malleability of our body representations?

Referred sensations: If the examiner pours cold water on the patient's face, the patient feels the sensation of cold both on her face and on her phantom hand (Ramachandran and Hirstein, 1998).

According to one of the dominant models, phantom sensations result from the cortical reorganization that follows amputation: inputs from the face invade, so to speak, the non-existent hand area in the primary somatosensory cortex (Pons et al., 1991).[1] Hence, when the face is touched, one can experience sensations in one's phantom hand. Should one say that amputees *mislocalize* their sensations? And should one compare their phantom sensations with referred sensations such as a pain from a heart attack that is felt in the arm? The crucial question here is to understand how one ascribes bodily sensations to the right part of the body. Does the representation of bodily space follow the same rules as the representation of external space? The pain that I feel in my hand is not felt in the cookie jar because my hand is in the cookie jar (Coburn, 1966). Why not?

Seeing the phantom: A mirror is placed vertically so that the mirror reflection of the intact hand is "superimposed" on the felt position of the phantom hand. Patients have then the illusory visual experience of their phantom hand through the 'mirror box', although what they

---

[1] For critical discussion, see Mezue and Makin (2017) and Medina and Coslett (2016).

actually see is their contralateral hand. Seeing as if their phantom hand were touched induces tactile sensations and seeing as if it were moving induces the illusion of phantom movements, even if the phantom hand has been paralyzed for ten years (Ramachandran and Rogers-Ramachandran, 1996).

When I close my eyes, I am still aware of my body: I feel it *from the inside* thanks to various bodily senses. The philosophical tradition has primarily focused on such an experiential mode in order to highlight how the awareness of our own body differs from the perception of other physical objects and of other people's bodies, all being accessible through external senses (Merleau-Ponty, 1945; Bermúdez, 1998). However, the dichotomy between the awareness of the body from the inside and from the outside may not be as clear-cut. The awareness of the body from the inside can indeed be influenced by its awareness from the outside. But is the role of vision merely anecdotal or does it reveal the fundamental multimodal nature of bodily awareness?

> *The rubber phantom*:   If the examiner simultaneously strokes the patient's stump and a visible rubber hand located in the continuity of the stump, the patient starts feeling as if the rubber hand were part of her body, as if it were her phantom hand (Ehrsson et al., 2008). A similar illusion of ownership, known as the rubber hand illusion, can be done in healthy individuals (Botvinick and Cohen, 1998). The participants already have two hands, and yet they report that it seems to them as if the rubber hand were part of their body.

Some amputees wear prostheses. Some of them conceive of their prostheses as mere tools, like glasses that they can put on and take off. Others regard them as intrinsic parts of their body like any other biological body part. What is the difference? To what extent can we embody external objects? What grounds the experience of the rubber hand or of the prosthesis as one's own? In the rubber hand illusion, it results from multisensory interaction. But does control over prostheses contribute as well? And what are the consequences of experiencing ownership, whether it is towards a rubber hand or towards one's own hand? Is there anything beyond a "feeling of warmth and intimacy"?

> *Mental amputation*:   After sensory or motor loss, or after brain lesion, one can feel as if a part of one's body were no longer one's own. For

instance, in somatoparaphrenia, patients deny that their hand is their own, despite still experiencing bodily sensations in it. Their feeling of confidence cannot be shaken and they can even attribute the so-called "alien" hand to another person. In another syndrome characterized by a sense of bodily disownership known as xenomelia (or Body Integrity Identity disorder), patients even wish their "alien" limb to be cut off.

The experience of an "alien" hand reveals what it is like to feel disembodied. In a sense, it is more startling than the experience of a phantom limb. As the saying goes, you only appreciate what you've had when it is gone. Disownership syndromes go against all our certainties about our own body. It is no longer James's "same old body" that those patients experience. It forces us to challenge the assumption that having a body and experiencing it as one's own always go hand in hand. What makes these patients experience their limbs as alien? What is missing? By contrast with patients with phantom limbs, patients with somatoparaphrenia can see their "alien" hand and they acknowledge that it is continuous and contiguous with the rest of their body that they experience as their own. Yet this does not suffice for them to feel that it is part of their body (see Appendix 3 for a detailed description). They can also feel when it is touched and pinpricked. But again, this does not seem to be sufficient to elicit the experience that they are touched on *their own hand*. This is so despite the biological fact that they receive tactile signals only from their own body and there should be no doubt that it is their own hand that is touched. Are these patients merely irrational? Is it because they can no longer control their hand? Or is there a fundamental difference between their bodily experiences and ours?

General Outline

The aim of *Mind the Body* is to provide a comprehensive treatment of bodily awareness and its underlying body representations, combining philosophical analysis with recent experimental results from cognitive science. To do so, I shall draw on research in philosophy, psychology, psychopathology, and cognitive neuroscience. In recent years, there has indeed been a renewed interest in body representations within cognitive science, revealing that there is a vast unexplored field beyond a constant blurry fuzzy bodily feeling. In particular, new bodily illusions, such as the rubber hand illusion, have raised a wide range of questions about the underlying mechanisms of the sense of bodily ownership. Drawing on the data derived from the study of bodily disruptions and bodily illusions (see Appendices 1 and 2 for a tentative glossary), I shall provide an account of the sensory, motor, and affective underpinnings of bodily awareness. I will start with the puzzles that any theory of the sense of bodily ownership faces (Part I), which require a better understanding of bodily experiences and bodily representations (Part II), before one might be able to solve them (Part III). The objective of *Mind the Body* is thus to provide for the first time a systematic conceptual framework aimed at promoting the integration of philosophical and scientific insights on bodily awareness.

PART I

Body Snatchers

"Merely—you are my own nose."
The Nose regarded the major and contracted its brows a little.
"My dear sir, you speak in error" was its reply. "I am just myself—
myself separately."

Gogol (1835)

We are aware of our bodily posture, of our temperature, of our physiological balance, of the pressure exerted on our skin, and so forth. Insofar as these properties are detected by a range of inner sensory receptors, one may conceive of bodily awareness on the model of perceptual awareness. However, there are other features of the body that are of a higher level and that cannot be directly extracted at the sensory level. They characterize what may be described as the fundamental state of the body, that is the enduring relation of the body with the world and with the self. We are aware that the body is here in the external world, that it belongs to us, and that it has two arms and two legs that can do some things but not others. Low-level bodily properties fluctuate all the time: you feel cold and then you do not; you are thirsty, you drink, you feel better until next time, and so forth. By contrast, the fundamental core of bodily awareness is relatively permanent, and thus rarely at the forefront of consciousness: it does not attract attention because it normally does not change. One may even ask: do we actually ever *feel* the body that way, or do we merely *know* those fundamental facts? And what is the origin of such awareness?

A good starting point is to compare the awareness of our own body with the awareness of other bodies. Through vision, audition, touch, smell and taste, we have access to bodily properties, whether they are instantiated by our body or by other bodies. But the classic five senses do not exhaust the list of ways of gaining information about our own body and there are also bodily senses, which give information about no other bodies than our own. Thanks to their privileged relation to our body, bodily experiences seem to afford not only awareness of our body *from the inside* but also awareness of our body *as being our own*. If so, the sense of bodily ownership requires no further analysis: feeling the posture of my leg is all it takes to be aware of this leg as being my own (e.g. Brewer, 1995; Cassam, 1997; Martin, 1992, 1993, 1995). Thanks to their privileged relation to our body, bodily experiences might also guarantee that I cannot be wrong about the body that I judge as my own: I cannot rationally doubt that this is my leg that I feel in an awkward posture.

However, the situation might not be as simple. Although most of the time, the body that we are aware of as our own is our own body, this is not always true and it is possible to report feeling our own body as foreign to us, and conversely, feeling an external body as our own. The objective of Part I is thus to reassess the relationship between bodily experiences and the sense of bodily ownership at both the phenomenological and the epistemological levels and to provide a roadmap for what lies ahead in our exploration of bodily awareness.

In Chapter 1, I will argue that there is something it is like to experience one's body as one's own. In Chapter 2, I will further argue that it takes more than feeling sensations as being located in a body part to have a sense of ownership. In Chapter 3, I will turn to the epistemological properties of the sense of ownership in relation both to bodily experiences and to visual experiences. I will argue that the sense of ownership does not require self-identification, but this does not entail that the lack of self-identification suffices for the sense of ownership.

1

Whose Body?

When I report that I feel my legs crossed, there are two occurrences of the first-person. The first occurrence refers to the subject of the proprioceptive experience (*I* feel), and it reveals the subjectivity of my bodily sensation (what it is like *for me*). The second occurrence of the first-person refers to the limbs that feel being crossed (*my* legs), and it reveals the sense of bodily ownership (the awareness of the legs that are crossed as being *my own*). Whereas it is the former occurrence of the first-person that has attracted most attention from philosophers, my focus will be on the latter. What is it like to feel one's limbs as one's own? But first, does it feel like anything?

Advocates of the liberal view reply positively. They claim that we have a primitive nonconceptual awareness of bodily ownership, which is over and above the experience of pressure, temperature, position, balance, movement, and so forth (Billon, 2017; Peacocke, 2014; Vignemont, 2013; Gallagher, 2017). However, according to a general principle of phenomenal parsimony, it may seem that one should avoid positing additional phenomenal properties into one's mental ontology as much as possible (Wu, forthcoming). If one grants that there are feelings of bodily ownership, there is indeed a risk of an unwarranted multiplication of feelings: feelings of a tanned body, feelings of a well-dressed body, feelings of a body in a train, and so forth. Advocates of the conservative view thus reject a distinctive experiential signature for the sense of bodily ownership: ownership is something that we believe in, and not something that we experience (Alsmith, 2015; Bermúdez, 2011, 2015; Wu, forthcoming). As summarized by Bermúdez (2011, p. 167):

There are facts about the phenomenology of bodily awareness (about position sense, movement sense, and interoception) and there are judgments of ownership, but there is no additional feeling of ownership.

The problem with the conservative view is that there is no clear guideline on how to apply the principle of phenomenal parsimony and one should not neglect the opposite risk of greatly impoverishing our mental ontology and denying phenomenology even when it is the most intuitive. The crucial question is thus to determine what reasons—if any—there are to posit feelings of bodily ownership.

This debate has recently turned to cognitive science to find answers, and more specifically to illusions and pathological disorders of bodily awareness. The appeal of these borderline cases can be easily explained. To some extent, they are simply thought experiments that happen to be actual.[1] Like thought experiments, they allow evaluating necessary and sufficient conditions, although only in our world and not in all possible worlds. The difficulty, however, is that there is room for interpretation in the analysis of these empirical cases and where the liberals see feelings, the conservatives see cognitive attitudes. The debate over the sense of bodily ownership has thus partly become a debate about how to best understand this syndrome or that illusion.

The objective of Chapter 1 is to provide a detailed description of the main empirical findings that have played a key role in recent discussions, including the rubber hand illusion (RHI) and syndromes of bodily disownership. I shall determine to what extent they can be taken as evidence for the existence of feelings of bodily ownership. Cognitive science may not always be able to adjudicate the debate among the different views but at least empirical cases constrain the shape that a theory of ownership should take by providing an agenda of what it has to explain.

1.1 A Phenomenal Contrast

How shall one settle the debate on ownership feelings? One way to proceed may be to apply the methodology used for other kinds of high-level properties, and more specifically the method of phenomenal contrast proposed by Siegel (2010). It proceeds in two steps. First, one describes a situation in which there is intuitively a phenomenal contrast

---

[1] As we shall see in Chapter 2, sometimes what is originally conceived of as a mere thought experiment can actually be found in neurology and psychiatry textbooks. Conceptual analysis ultimately rests on intuitions but sometimes empirical facts contradict one's intuitions.

between two experiences, one of which only instantiating the high-level property. For instance, it has been suggested that it does not feel the same when one hears a sentence in a foreign language before and after learning to speak the language. Other typical examples are seeing a succession of events with or without perceiving their causal relation (Siegel, 2009), and seeing something as a pine tree before and after becoming an expert spotter of pine trees (Siegel, 2010). The second step consists in drawing an inference to the best explanation of this contrast by ruling out alternative interpretations. Most interest in the debate on the degree of richness of perceptual content has focused on visual awareness but it may also be applied to bodily awareness. The only difference here is that it is far more difficult to find scenarios in which one is not aware that this is one's own hand than scenarios in which one is not aware that this is a pine tree. Nonetheless, there are some borderline cases in which one can manipulate the awareness of the ownership property, which I shall now describe.

1.1.1 The Rubber Hand Illusion

The RHI has become the goose that lays the golden eggs for those interested in the sense of bodily ownership. In the classic experimental set-up, one sits with one's arm hidden behind a screen, while fixating on a rubber hand presented in one's bodily alignment; the rubber hand can then be touched either in synchrony or in asynchrony with one's hand. After a couple of minutes, it has been repeatedly shown the following effects during the synchronous conditions only:

At the phenomenological level (measured by questionnaires):

- *Referred tactile sensations*: Participants report that they feel tactile sensations as being located not on their real hand that is stroked but on the rubber hand.
- *Sense of ownership*: They report that it seems to them as if the rubber hand were their own hand.

At the behavioural level:

- *Proprioceptive drift*: They mislocalize the finger that was touched in the direction of the location of the finger of the rubber hand.[2]

---

[2] Although for a long time the proprioceptive drift was the main implicit measure of the RHI, it has been found that it can be dissociated from the explicit measure of

At the physiological level:

- *Arousal*: When they see the rubber hand threatened, they display an increased affective response (as measured by their skin conductance response).

Since the discovery of the RHI by Botvinick and Cohen in 1998, several other bodily illusions manipulating the sense of ownership have been found, including a somatic version of the RHI (Ehrsson et al., 2005), the invisible hand illusion (Guterstam et al., 2013), the supernumerary hand illusion (Guterstam et al., 2011), out-of-body illusions (Lenggenhager et al., 2007; Ehrsson, 2007), the body-swapping illusion (Petkova and Ehrsson, 2008), and the enfacement illusion (Tsakiris, 2008; Sforza et al., 2010) among others (see Appendix 1 for a list of bodily illusions). Some of these illusions use fake bodies while others appeal to virtual avatars or to another person's body. They can be local, restricted to the hand or to the face, or global, including the whole body. They can manipulate first-person or third-person perspective on the body. Despite these differences, in all of them participants report that it seems to them as if an extraneous body part were their own.[3]

There is clearly a contrast between the synchronous and the asynchronous conditions in the RHI but is it a *phenomenal* contrast? Does it *feel* different when the subjects report that the rubber hand is their own hand and when they do not? And if it does, is it because of a phenomenal property of ownership? Since the RHI is now conceived as *the* experimental paradigm to study ownership, it is especially important to examine what it exactly involves. The first question that one needs to answer is whether the RHI concerns the sense of bodily ownership at all. Some might indeed claim that the participants' ownership statements are a mere way of speaking, which is only loosely connected to the kind of awareness that they normally have of their own biological hands. If this were true, then one would not be entitled to draw conclusions about the sense of bodily ownership on the basis of this illusion and all the studies using this experimental paradigm would simply be missing their target.

---

ownership as rated by questionnaires (Rohde et al., 2011; Holmes et al., 2006; Abdulkarim and Ehrsson, 2016).

[3] In this book I shall not discuss full-body illusions and focus exclusively on the sense of ownership of body parts.

But what could be used as arguments in favour of such a radical conception? I will now list a number of empirical facts, on the basis of which one may be tempted to question the relevance of the RHI, but, as I shall argue, none of them constitutes fatal objections.

Let us first consider the introspective measure. Participants are asked to rate to what extent they agree with ownership statements such as "it seems to me as if the rubber hand belonged to me", "it seems to me as if the rubber hand were part of my body", and "it seems to me as if the rubber hand were my own hand". Their reply is positive but one can note the weakness of their approval: in one of the rare large-scale studies done with 131 participants, the mean response for ownership is only +0.4 after synchronous stroking on a scale that goes from −3 (strongly disagree) to +3 (strongly agree) (Longo et al., 2008). Surprisingly, in the same study they also only *weakly disagree* after asynchronous stroking when asked if it seemed as if the rubber hand were their hand (mean response −1.2 with a small standard deviation). These low ratings cast doubt on what the questionnaire really reveals. Under normal circumstances, if we were asked whether it seems to us that our hand belongs to us, we would most certainly give a strong positive reply while we would give a strong negative reply for a rubber hand. But what do these differences between experimental and normal situations amount to? Do they imply that the participants who undergo the illusion do not feel ownership? Or do they simply reveal different degrees of *confidence*? Even when the rubber hand feels like their own hand, the participants have many reasons to question what they feel and there is room for doubt. This does not show that the RHI does not manipulate the sense of ownership at all, but only that it does so imperfectly.[4]

Let us now consider differences at the behavioural and neural levels. Participants mislocalize their own hand but the proprioceptive drift is only partial: they mislocalize it *merely towards* the location of the rubber hand, and not at its exact location. Hence, one might conclude, they do not really take the rubber hand to be their own hand. However, one must be careful in one's analysis of the proprioceptive drift. Participants do not localize their hand at the actual location of the rubber hand, but the fact is that they do not localize the rubber hand at its actual location either.

---

[4] We shall return to the issue of confidence in Chapter 9.

Instead, its perceived location also drifts in the direction of the location of the participants' hand (Fuchs et al., 2016). Roughly speaking, they meet halfway, not so far from each other. It has also been found that there are different brain activities when participants feel touch to be located on their real hand and on the rubber hand: in ventromedial prefrontal cortex, in lateral occipito-temporal cortex, and in the temporo-parietal cortex (Limanowski and Blankenburg, 2016). But these differences in brain activities merely indicate that one experiences *referred* sensations on the rubber hand, which involve distinct mechanisms from sensations in one's hand. They have nothing to do with ownership. Therefore, there is no clear reason to discount the RHI as being relevant and this illusion should definitely be taken into account to illuminate our understanding of the sense of bodily ownership. But we should not restrict ourselves to it and we must also analyse the mirror phenomenon, when one is no longer aware of one's body as being one's own.

1.1.2 A Sense of Disownership

The more I gazed at that cylinder of chalk, the more alien and incomprehensible it appeared to me. I could no longer feel it as mine, as part of me. It seemed to bear no relation whatever to me. It was absolutely not-me—and yet, impossibly, it was attached to me—and even more impossibly, continuous with me.

(Sacks, 1984, pp. 47–8)

Oliver Sacks's (1984) vividly describes what it feels like when one has lost a sense of ownership for one's own limbs. His experience might be hard to conceive but we can have a glimpse of what it feels like when we fall asleep on our arm: when we wake up, our arm feels numb and almost as an alien dead object attached to our body. Whereas it suffices for us to change our position to feel the hot tingle of blood rushing back in along with a sense of ownership, there is a variety of what may be called disownership syndromes that can last for days or even weeks (see Table 1.1).

Sensory loss or motor loss, for instance, can induce a sense of bodily disownership. Suppose you have broken your leg and have worn a cast for several weeks. Afterwards, despite apparently complete recovery, you still barely use it (i.e. segmental exclusion syndrome). It is as if the limb, which has been out of sight and out of action for several weeks, had been excluded from the representation of your body and you need to relearn that it is part of it. The consequences of peripheral sensory and motor loss can be even more drastic after spinal cord injury. Patients then sometimes describe that their body appears "as a hole" as soon as they

Table 1.1. A sense of disownership for one's body

Denial of bodily ownership

Complex regional pain syndrome	Chronic pain syndrome often after injury, which causes intense burning, decreased ability to move the affected body part, and disownership.
Congenital pain insensitivity	Dramatic impairment of pain since birth, caused by a hereditary neuropathy or channelopathy, leading to a complete lack of discomfort or withdrawal reaction to noxious stimuli and to self-injuries.
Depersonalization	General alteration of the relation to the self: anomalous bodily experiences, emotional numbing, sensation of alienation from surroundings, and anomalous subjective recall.
Peripheral deafferentation	Loss of tactile and proprioceptive information.
Segmental exclusion syndrome	Underutilization of the limb after recovery following the lack of use of the limb caused by a traumatic or infectious affliction.
Somatoparaphrenia	Denial of ownership of one's body part.
Trauma in war prisoners	Sense of disembodiment as a result of confinement and isolation.
Xenomelia (or body identity integrity disorder)	Desire to have one's perfectly healthy limb(s) amputated.

close their eyes (Berger and Gerstenbrand, 1981; Head, 1920; Scandola et al., 2017). A similar type of experience can be found in patients with peripheral deafferentation after some very rare acute sensory neuropathy. One such patient, Ian Waterman, has no proprioception and no touch below the neck: if he closes his eyes, he does not know where his limbs are. Consequently, before he had learned how to exploit visual information about his bodily posture, he could barely control his limbs. He then reported feeling alienated from his body (Cole, 1995).

Another condition in which abnormal bodily awareness includes a sense of disownership is the psychiatric disorder of depersonalization. Patients with depersonalization have bodily sensations but they experience abnormal bodily properties (for instance, the size of their body parts feels distorted), and they feel detached from them, as if they were external observers of their mental and bodily processes. Close to 70 per cent of the patients with depersonalization feel as if their body did not belong to them and as if it had disappeared:

My body didn't feel like my own. I just felt like a floating mind in space with blunted, blurry thoughts. I found it crazy sometimes that I even had my own body, because I didn't feel connected to it. I'd look down at my hands and they didn't feel real. (Bradshaw 2016)

Another puzzling disownership syndrome is called xenomelia, but it is also known as apotemnophilia and body integrity identity disorder (for a review, see Sedda and Bottini, 2014). Patients with xenomelia have apparently normal sensory and motor functions. Yet they have an overwhelming desire to be amputated of one of their perfectly healthy limbs, often since childhood or early adolescence (First, 2005). The undesired limb is not perceived differently from the other limbs: it neither looks ugly nor deformed, nor does it feel impaired (Braam et al., 2006). The most frequent explanation offered by the patients is that their limb is superfluous and that it does not feel like part of their body (First, 2005; Braam et al., 2006; Hilti et al., 2013):

I don't understand where it comes from or what it is. I just don't want legs. Inside I feel that my legs don't belong to me, they shouldn't be there. At best my legs seem extraneous. I would almost say as if they're not part of me although I feel them, I see them, I know they are . . .

(Corrine in *Complete Obsession*, BBC, 17 February 2000)

The desire to be amputated may thus reveal a mismatch between the biological body and its representation (First, 2005; Bayne and Levy, 2005; Ramachandran and McGeoch, 2007; Hilti et al., 2013). With the help of surgical amputation, individuals with xenomelia hope to solve the conflict by aligning their biological body with their body representation. Some patients actually perform self-amputation, going as far as lying under a train, building a home-made guillotine, or freezing their legs in dry ice. When surgeons agree to cut off their undesired limb, they feel relieved and "complete", whereas they felt "incomplete" before the amputation.[5]

---

[5] However, further fixation with amputation of other limbs can occur after the initial satisfaction over the years. For instance, after the amputation of his leg above the knee, a patient self-amputated his right little finger, then his left little finger, and then his left ring finger, to finally cut his left forearm (Sorene et al., 2006).

There is a final neurological disorder of bodily disownership, known as somatoparaphrenia (or asomatognosia or alien hand sign), which I shall focus on throughout the book. It is caused by a lesion of or an epileptic seizure in the right parietal lobe and it is often associated with motor and somatosensory deficits and spatial neglect, but the main symptom is the fact that patients deny that one of their limbs belongs to them (Moro et al., 2004; Vallar and Ronchi, 2009) (see Appendix 3 for a detailed phenomenological description):

E: If this hand is not yours can I take it away with me? P: Of course! If you want it, I will give it to you as my gift, since I have no need for it. E: Do you want to move this hand away? Wouldn't you be sad without it? P: Yes, if it was mine, but it's not.
(Invernizzi et al., 2012, p. 148)

Furthermore, somatoparaphrenic patients attribute their so-called "alien" limb to another individual.

"It's my niece's hand". "She is so kind, she left it here to keep me company... she's also very absent minded...look here, she was rushing home and forgot her hand here"   (Gandola et al., 2012, p. 1177)

Interestingly, immediately after the stimulation of the nerve in the inner ear that provides the sense of balance (i.e. vestibular stimulation), somatoparaphrenic patients can momentarily regain a sense of ownership towards their hand before denying again that it belongs to them (Bisiach et al., 1991; Rode et al. 1992):

Ex: Whose arm is this?
A.R.: It's not mine [. . .] Feel it's warmer than mine.
Ex: So where is *your* left arm?
A.R.: (*Makes an indefinite gesture forwards.*) It's under there.
Immediately after vestibular stimulation, the examiner asks the patient to show her the patient's left arm.
A.R.: (Points to her own left arm.) Here it is.
Ex: (Raises the patient's left arm.) Is this arm yours?
A.R : Why, yes.
[. . .] *Two hours after vestibular stimulation A.R. is questioned again by the examiner.*
Ex: (Points to the patient's felt arm.) Whose arm is this?
A.R.: It's my mother's, It's warmer.
Ex: Where IS your left arm?
A.R. Stares silently at the examiner (Bisiach et al. 1991, p. 1030).

There is clearly a contrast between before and after vestibular stimulation in somatoparaphrenia, but is it because of a phenomenal property of ownership or only because of a phenomenal property of *dis-ownership*? In other words, does the phenomenology of disownership merely express the absence of the phenomenology of ownership? Baier and Karnath (2008) suggest using the term of "disturbed sensation of limb ownership" to refer to somatoparaphrenia-like syndromes, thus implying that there is normally a sensation of ownership that is missing in these patients. This analysis fits the patients' descriptions: they seem to report more that their limb does not feel like being their own ("I don't feel it as my hand" Cogliano et al., 2012, p. 764) than that it feels like being alien. However, this may not convince those who think that one should not take the patients at their word and who question the relevance of these pathological disorders for our understanding of the sense of bodily ownership. On their view, these patients experience a feeling of disownership but they do not lack the feeling of ownership that healthy subjects normally experience because there is no such thing (Bermúdez, 2015).

It is true that disownership syndromes on their own do not suffice to prove the existence of ownership feelings. However, if we have independent evidence that they normally exist and that by default there is a phenomenology of ownership, then it is highly likely that it is impaired in the patients who suffer from these syndromes. Still the experience of disownership does not simply equate to the lack of experience of ownership. Like most bodily awareness, the phenomenology of ownership is recessive, staying at the margin of consciousness. As Frith (2005, p. 767) said: "Paradoxically it seems as if the mark of the self in action is that we have very little experience of it". Likewise, we have very little experience of bodily self. Although we receive a constant flow of information about our body, we are most of the time barely aware of our body, let alone of our body as our own. The fact is that our body never leaves us and it can thus become almost transparent. It is like the painting that you have always seen on the wall in your parents' house: you no longer see it. And if one day the painting is gone, you may not notice it because you have taken it for granted, so to speak. Your visual experience of the painting is then absent in your visual phenomenology but you do not have the experience of its absence. Likewise the absence of the experience of ownership can go unnoticed. In such cases, I would argue, one does not experience disownership. When one does experience disownership, it is not only because of the absence of the experience of ownership; it is also because one has an experience of its absence.[6]

1.2 A Cognitive Interpretation

Some conservatives might grant that the sense of ownership is altered in the RHI and in disownership syndromes, but still dispute that the reports that the subjects and the patients make reveal anything more than *judgements* of ownership (Wu, forthcoming). These reports are indeed open to interpretation and they may express doxastic states instead of phenomenal ones. If so, we should not take the subjects' reports at face value when they claim that they *feel* that the rubber hand is their own or when they claim that they *feel* that the hand does not belong to them. As Anscombe (1957, 1962) notes, the use of the term "sensation" can be misleading: it is not because we talk of sensations of a specific feature that we do have them. Instead, what we may really have are only beliefs or judgements about this feature. She famously concludes that we do not *feel* our legs as being crossed; we simply *know* that they are that way. Likewise the conservatives may claim that the subjects merely *judge* that it is their own hand or that it is not. The main challenge is thus to decide whether we have "true" sensations or not. We shall see that in the case of bodily ownership there are no empirical or a priori reasons to favour the cognitive interpretation over the experiential one.

1.2.1 What Cognitive State?

The cognitive interpretation has some plausibility as far as somato-paraphrenia is concerned because this pathology is characterized by delusional beliefs, which are of two types:

(i) Disownership delusions: they believe that this body part does not belong to them.
(ii) Confabulatory delusions: either they attribute the limb to another individual or they personify it.

---

[6] Levine et al. (1991) defend a similar view for anosognosia for hemiplegia (see also Ramachandran, 1995). They argue that sensorimotor deficits are not phenomenologically salient and need to be discovered. In other words, one assumes that one's body is healthy unless one is provided with evidence to the contrary.

For both types of delusions, the patients display an unshakable feeling of confidence: "Feinberg: Suppose I told you this was your hand? Mirna: I wouldn't believe you" (Feinberg et al., 2005, p. 104). However, according to the current most influential theory of delusion, the two-factor model, the thematic content of a particular delusion finds its origin in sensory or motor impairment leading to abnormal *experiences* that the patient tries to account for (Langdon and Coltheart, 2000). This seems to be confirmed in somatoparaphrenia, in which the disownership delusion appears to be strongly anchored in abnormal feelings. A patient, for instance, reported:

my eyes and my feelings don't agree, and I must believe my feelings. I know they look like mine, but I can feel they are not, and I can't believe my eyes.

(Nielsen, 1938, p. 555)

Furthermore, there are other cases in which it simply seems to the patients *as if* it were not their hand, although they do believe that it is their own hand. For instance, patients with depersonalization are not delusional: they know that this is their own body. A patient thus reported: "I can sit looking at my foot or my hand and not feel like they are mine" (Sierra, 2009, p. 27). She correctly self-ascribed her foot and her hand but still experienced as if they were not hers. Likewise, participants in the RHI are fully aware that the rubber hand is a fake hand, a mere piece of rubber. It does not seem to be likely that they judge at the same time both that the rubber hand is their own hand and that it is not. Instead, what they experience appears as equivalent to what subjects experience in classic perceptual illusions, such as the Müller-Lyer one, in which the visual experience is not affected by the belief that the two lines are of the same size. It may also be compared to what one experiences when one feels vertigo: one knows that there is no actual risk of falling down and yet one feels afraid to fall. In these perceptual and emotional cases the subject is in an experiential state with a distinctive phenomenology whose content is at odds with the content of her belief. The liberal conception proposes a similar analysis for the ownership cases: in the RHI subjects cannot help but *experience* the rubber hand as being part of their own body, whereas depersonalized patients cannot help but *experience* their hand as not belonging to them.

The liberal conception thus appeals to dissociations of these sorts between what it seems to the subject and what the subject believes to argue for the experiential dimension of ownership. However, this argument schema, which has been called the argument from cognitive impenetrability (Mylopoulos, 2015), has been criticized on the ground that attitudes other than feelings and sensations can be encapsulated and immune to the influence of beliefs and judgements (McDowell, 2011). But if the patients and the subjects in the RHI experiments do not express what they feel, then what kind of state do they express?

Wu (forthcoming) proposes that participants in the RHI report "a gradable attitude of agreement towards a proposition". To some extent Wu's proposal is hardly controversial: level of agreement is precisely what is asked for in the RHI questionnaires. But what grounds the participants' replies? According to Wu, participants draw an inference on the basis of evidence that points to features that they would expect if the rubber hand were their own hand. The problem, however, is that there can be strong defeaters, including the fact that they already have two hands and that there are differences in skin colour, size, and laterality between the rubber hand and the real hand. Yet they do not cancel the illusion (Longo et al., 2009a; Petkova and Ehrsson, 2009). How can participants even partly agree that this is their own hand when it does not look at all like their hand? To counterbalance the defeaters there must be strong evidence that this is their own hand and what more powerful evidence than the participants' feelings?

Alternatively, proponents of the cognitive interpretation may go for something weaker than partial agreement and suggest that participants only *suppose* that the hand is their own while they know that this is not true. As far as this attitude is not driven by evidence, it does not matter whether there are strong defeaters. However, it does not seem plausible that the patient suffering from depersonalization simply *supposes* her hand not to be her own, while believing it is. One problem indeed is that she is not in a cold cognitive state about the lack of bodily ownership. Her introspective reports are associated with strong distress. Why are depersonalized patients greatly disturbed? They know that their body is still their own, it only seems to them as if it were not. How can mere suppositions have such an emotional impact?

Imagination may then be a better candidate because it can be affectively loaded. One may suggest that the depersonalized patient *imagines* her body not to be her own and that imagining suffices to make her feel upset. Along the same lines, Alsmith (2015) argues that in the RHI one imagines that the rubber hand is one's hand although one believes that it is not. However, imagination is generally under voluntary control whereas depersonalized patients cannot stop feeling the disappearance of their body. Nor can subjects in the RHI voluntarily decide whether they are susceptible to the illusion or not. Even if imagination can escape control, the reason for which it can be affectively loaded is that it includes a phenomenal component. On one interpretation indeed, imagination involves mentally recreating selected experiences (Goldman, 2006). If one does imagine the rubber hand as being one's own, it means that one imagines *experiencing* it as one's own. The imaginative proposal is thus not purely cognitive; it includes the experience that is re-enacted in imagination.

This brings us to a more general remark. One should clearly distinguish the following two questions: (i) is there a phenomenology of ownership? And (ii) is the sense of ownership cognitive? A positive reply to the second question does not entail a negative answer to the first one. Advocates of cognitive phenomenology might indeed argue that even if bodily ownership could be represented only conceptually, it would not preclude it from being experienced. For instance, it has been recently suggested that we have a range of noetic feelings (Dokic, 2012). Suppose I have a déjà vu experience although I perfectly know that I have never been here before. In this case, it seems legitimate to assume that it feels something specific. Arguably, this feeling is not sensory and its grounds are relatively complex, involving metacognitive monitoring of visual processing. One can also mention the feeling of familiarity: it is characterized by a specific phenomenology that goes beyond sensory processing and that can be conceived of in affective terms (Dokic and Martin, 2015). If we accept that there are feelings of familiarity, déjà vu, and so forth, then why not feelings of ownership? Ownership feelings might then be conceived of in the same way as these non-sensory types of feelings. Some of those who claim to reject the liberal conception actually reject only the hypothesis of a *sensory* phenomenology of ownership and may be willing to accept such a cognitive view (e.g. Alsmith, 2015). Their view still differs from a purely liberal conception because they do not assume that the ownership judgement is grounded in an ownership experience. Instead, the ownership judgement *gives rise* to an ownership experience.

To recapitulate, proponents of the conservative conception avoid the burden to posit a phenomenological quality of ownership but it is unclear how their cognitive interpretation can account for the borderline cases of ownership that we have described here. Indeed the alleged cognitive attitude that the conservatives postulate must be not only insensitive to the influence of beliefs; it must also be beyond one's control and affectively loaded. So far imaginative states may be the best candidate but only because they consist in experiential states. Therefore the argument from cognitive impenetrability may not be a sufficient proof for the existence of ownership feelings but it is still a powerful one. Each of the borderline cases might be open to interpretation but taken all together they provide strong support for a liberal conception of bodily phenomenology, which can provide a simple and unified explanation. However, for an argument to the best explanation to work here one needs to make sure that there are no fundamental objections against the liberal interpretation. I shall now focus on Anscombe's original epistemological argument.

1.2.2 Bodily Knowledge without Observation

A man usually knows the position of his limbs without observation. It is without observation, because nothing shews him the position of his limbs; it is not as if he were going by a tingle in his knee, which is the sign that it is bent and not straight. Where we can speak of separately describable sensations, having which is in some sense our criterion for saying something, then we can speak of observing that thing.

(Anscombe, 1957, p. 13)

According to Anscombe (1957, 1962), whatever we may feel about our body, it cannot explain what we know about it. Bermúdez (2015) further adds that if bodily feelings have no epistemic role, then there is no reason for them to exist. Our bodily knowledge is then what Anscombe calls "knowledge without observation". One may, however, wonder why bodily feelings cannot ground knowledge. Anscombe replies that only sensations whose internal content is "separately describable" can play such a role. On her view, there is a sensation of x if its description has a different content than x and this content is taken as a sign that indicates x. For example, there is a sensation of going down in a lift since one can provide an independent description of its internal content in terms of lightness and of one's stomach lurching upward. By contrast, it is not legitimate to talk of a sensation of sitting cross-legged, Anscombe claims, because one cannot give such a separate description. We can reconstruct her argument as follows:

(i) Only content that is separately describable can be used as the epistemic basis for judgements.

(ii) A state qualifies as a sensation only if its content can be described independently of what the sensation refers to.

(iii) One cannot provide an independent description in the case of bodily posture.

(iv) Thus, there are no sensations of bodily posture.

(v) Thus, bodily knowledge is knowledge without observation.

Along the same lines, Bermúdez (2017) argues that ownership feelings are a mere "philosophical fiction". On his view, sensations that can be used as epistemic basis must meet two criteria (Bermúdez, 2015). On the one hand, they must be "focused": their content must provide information that is precise and specific. If they are too vague, they fail to justify the specific judgement that one makes. On the other hand, they must be independent: they cannot simply duplicate the content of the judgement because, Bermúdez (2015) claims, one cannot justify an assertion by simply repeating it. He then concludes that the feeling of ownership fails to meet these two criteria.

> And yet a feeling of myness that can only be described in those very terms is not sufficiently independent of the judgment of ownership that it is claimed to justify. So the postulated non-conceptual intuitive awareness of ownership falls foul of Anscombe's dilemma. (Bermúdez, 2015, p. 39)

On Anscombe's and Bermúdez's view, we can decide whether or not there are bodily feelings by determining the epistemic ground of bodily knowledge. However, the notion of knowledge without observation and its application to bodily knowledge have given rise to a large debate, partly because of its obscurity (e.g. Moran, 2004; Schwenkler, 2015). What is surprising is that they reach their metaphysical conclusion on the basis of epistemological considerations. One may thus be tempted to immediately rule out their argument because it assumes the priority of epistemology over metaphysics. For many indeed, whether or not there are feelings is a matter independent of epistemic issues. But even if one accepts Anscombe's epistemology-first strategy, one may still challenge her specific epistemological theory: why do contents have to be separately describable to ground judgements? And how to characterize such contents? The distinction between the sensations that can be separately describable and those that cannot is problematic. Anscombe's (1962, p. 57) own examples are not helpful. For instance, she claims that "the visual impression of a blue expanse" is independent and can thus ground the judgement that the sky is blue. Since she defends the view that sensations can be captured in intentional terms (Anscombe, 1965), her point is not that there is a qualitative raw feel of blueness independently of the visual property of blue. But then it is not clear in what sense "the visual impression of a blue expanse" qualifies as an independent description. It is thus sometimes difficult to see why bodily awareness and perceptual awareness do not fall into the same category.

Finally, the whole argument rests on the assumption (iii) according to which one cannot provide an independent description in the case of bodily position: "no question of any appearance of the position to me, of any sensations which give me the position" (Anscombe, 1962, p. 58). Similarly, Bermúdez (2015, p. 39) assumes that one cannot describe the sense of ownership without referring to the fact that this is one's own body (i.e. myness):

It is highly *implausible* that there is a determinate quale of ownership that can be identified, described and considered independently of the myness that it is supposed to be communicating.   (Bermúdez, 2015, p. 39, my italics)

But why is it so? Neither Anscombe nor Bermúdez give arguments. They merely state that there cannot be independent descriptions. Bermúdez (2015, p. 44) further adds that ownership is "a phenomenological given" and that it is impossible to ground it in further nonconceptual content. But this seems to be simply begging the question. By offering no principled reasons for their third assumption, they undermine their own objection.

1.2.3 Beyond Myness

At this point, one may note that Bermúdez interchangeably uses the terms of feeling of ownership and feeling of *myness*. However, one may wonder whether they capture exactly the same notion.[7] Feelings of myness (or mineness) have recently attracted a lot of attention in the literature on consciousness (Guillot, 2017; Guillot and Garcia-Carpintero, forthcoming). Most discussion has focused on the subjective quality of conscious states

---

[7] This makes Bermúdez's view sometimes difficult to understand. For example, he explicitly rejects a "positive phenomenology of ownership" (2011, p. 163), but he also claims, "there is a phenomenology of ownership" (2015, p. 38).

(i.e. what it is like for me to be in this state, cf. Kriegel, 2009) or on mental ownership (i.e. I experience this sensation as of being my own, cf. Lane, 2012). But one might also apply it to the body that we experience as our own. Although the notion of myness is most of the time left undefined, we can give the following definition in representational terms: an experience of bodily ownership consists in a feeling of myness if its content includes the relation of ownership (e.g. "as of being my own"). Such a feeling may seem attractive because of its apparent explanatory power for the first-personal character of the sense of bodily ownership. Any theory of the sense of bodily ownership must indeed account for the fact that the phenomenology of bodily ownership normally grounds self-ascriptive judgements (e.g. this is my own body). It may then seem that the simplest way to account for this first-personal character is to appeal to feelings of myness: an experience of bodily ownership with a content of the type "this is mine" seems indeed to be a legitimate ground for bodily self-ascriptions.

We have seen that Bermúdez denies that the feeling of myness can play such a role because it would not be independent enough. But even if his argument about myness was valid, would that imply that the liberal conception is false and that there could be no feelings of ownership? The reply is negative because there may be ways to spell out the content of ownership feelings other than in terms of myness, which may then be separately describable. Let us try a more modest definition: the feeling of ownership is a distinctive phenomenological quality in virtue of which one is aware of one's body as one's own. One type of such quality is the feeling of myness, which consists in a feeling that is *explicitly about* the fact that the body is one's own. This specific feeling is Bermúdez's actual target. But as we shall see in Part III, there may be alternative liberal versions that do not represent ownership as such, which can still account for the first-personal character of the sense of bodily ownership.

1.3 Conclusion

We started this chapter with two contrasts between situations in which one is aware that a hand is one's own and situations in which one is not:

(i) *The rubber hand illusion*: it seems to one that the rubber hand is one's own hand in the synchronous condition, but not in the asynchronous one.

(ii) *Disowership syndromes*: it seems to one that one's hand is not one's own before vestibular stimulation, but not after.

We saw that these two contrasts could not be easily accounted for in purely cognitive terms. According to the liberal conception, it makes a phenomenological difference when one is aware of one's body as one's own and when one is not. As described by Peacocke (2014):

A hand's being experienced as yours is part of the phenomenology of ordinary human experience. As elsewhere, this phenomenology should not be identified with any kind of judgement of a content "that's mine". That judgement is not made (in fact, it is rejected) when a subject knows that he is experiencing the rubber hand illusion, but he still experiences the rubber hand as his.

(Peacocke, 2014, p. 51)

Now it is time to determine what phenomenological properties explain the phenomenal contrast: it may be ownership feelings but it may also be other features of bodily awareness. Some versions of the conservative conception can indeed grant that there is a phenomenal contrast between cases in which one is aware of one's hand as one's own and cases in which one is not, but explain the contrast in terms of the phenomenology of other sensory features of the body. Let us go back to the example of the RHI. Clearly this illusion is not only about bodily ownership. First, participants mislocalize their hand in the direction of the location of the rubber hand. Secondly, they experience referred sensations in the rubber hand: they feel the stroking of the paintbrush as of being located on the rubber hand instead of their own hand. These proprioceptive and tactile effects occur only after synchronous stroking when subjects report ownership. One may then argue that these proprioceptive and tactile differences exhaust the phenomenal contrast. More generally, one may defend the view that the felt location of bodily sensations suffices for the sense of bodily ownership. In Chapter 2 we shall consider in detail this proposal and argue that the phenomenology of ownership is over and above bodily sensations.

2

Over and Above Bodily Sensations

When one feels a bodily sensation to have a location there is no issue over whose body it appears to belong to (see O'Shaughnessy 1980, volume 1, p. 162). Rather in as much as it feels to have a location, it feels to be within one's own body.

(Martin, 1992, p. 201)

According to Martin, one feels as one's own the limb in which one feels sensations. Nothing more is required for the phenomenological quality of ownership because it consists only in the spatial phenomenology of bodily experiences. Hence, it makes a phenomenological difference when one is aware of one's body as one's own and when one is not but this difference is purely spatial; it is not about ownership as such. As Martin (1992, pp. 201–2) claims, "What marks out a felt limb as one's own is not some special quality that it has". There is nothing over and above the location of the sensations.

But is the sense of bodily ownership that simple? It is true that one normally experiences ownership when one feels bodily sensations but is it sufficient to feel sensations as being located in a part of one's body to experience this body part as one's own? As we shall see, one can have bodily sensations with no sense of ownership, and even with a sense of disownership. After describing Martin's view in more detail, I will present a series of counterexamples and argue that his view fails to account for the first-personal character of the sense of bodily ownership.

2.1 The Sole-Object View

Let us imagine that I wake up in the middle of the night. I feel my heart beating too fast. The position of my left arm feels uncomfortable and painful. It feels nice to stretch on the bed and to feel the contact of the cold sheet on my skin. I feel too warm. I am thirsty. I get up but I lose my balance. These various types of information are directly available only about my own body. One may then claim with Locke (1689) that by experiencing what affects my limbs I experience them as being parts of me.

> That this is so, we have some kind of evidence in our very bodies, all whose particles, whilst vitally united to this same thinking conscious self, so that we *feel* when they are touched, and are affected by, and are conscious of good or harm that happens to them, are a part of our *selves*, i.e. of our thinking conscious *self*. Thus the limbs of his body are to everyone a part of himself; he sympathizes and is concerned for them. Cut off an hand, and thereby separate it from that consciousness he had of its heat, cold, and other affections; and it is then no longer a part of that which is *himself*, any more than the remotest part of matter.
>
> (Locke, 1689/1997, bk. II, ch. 27, §11)

What we want to know, however, is not how the limbs are *parts of oneself* or how one is aware of it, but how one is aware of the limbs *as parts of one's body*. Whether one's body is oneself or not, that is another issue. Still one may propose that the fact that we experience from the inside what affects our body is also the key to the sense of bodily ownership.

Individuating one's own body involves being able to discriminate it from what is not one's body. To avoid a circular account of ownership, this discrimination cannot be phrased in terms of my body versus not my body. The elegant solution proposed by Martin is to phrase it in spatial terms, between inside and outside bodily boundaries in which one can experience bodily sensations. The sense of ownership then involves being aware that one's body has limits, that it has spatial boundaries beyond which one cannot feel sensations.

One may reply that one can be aware of the boundaries of the body without being aware of the boundaries of the body *qua* one's own. This is the case, for instance, in visual experiences: I can be visually aware of my body as a bounded object within a larger space when I see it. However, there is a fundamental difference in spatial organization between visual experiences and bodily experiences. Consider first the case of vision. The boundaries of the body that I see are not co-extensive with my visual field. Many other objects can occupy my visual field, possibly including many other bodies. Therefore, visual awareness of bodily boundaries cannot confer a sense of ownership on the body part that is seen (Brewer, 1995; Bain, 2003; Martin, 1995). By contrast, when I have a bodily sensation, I do not feel it in one body as opposed to another body. According to what Martin (1995) calls the sole-object view, in order for an instance of bodily experience to count as an instance of perception, it must be an experience of what is in fact the subject's actual body. Bodily awareness has a unique object, namely, one's own body, and there cannot be veridical bodily experiences that do not fall within the limits of one's own body. There is thus an identity between one's own body and the body in which one locates bodily experiences and this identity, Martin claims, enables the spatial content of bodily experiences to ground the sense of bodily ownership.

Sensations that are felt beyond the actual limits of one's body, as in the case of phantom limbs, are not problematic for Martin's view because the phantom limbs are experienced within the apparent confines of the bodies that the amputees experience as their own. The same can be said of the RHI. One way to interpret the illusion is indeed that one experiences the rubber hand as one's own in virtue of feeling tactile sensations as being located in it. These sensations, on Martin's view, should be conceived as hallucinatory or illusory but they validate the hypothesis that it suffices to feel sensations as being located in a body part or in an object to experience this body part or this object as one's own: "this sense of ownership, in being possessed by all located sensations, cannot be independent of the spatial content of the sensation, the location of the event" (Martin, 1995, p. 277).

What would be more problematic for Martin is if one could feel sensations in a limb and yet fail to experience the limb as being one's own. But Martin assumes that this is simply impossible:

> If the sense of ownership is a positive quality over and above the felt quality of the sensation and the location—that there is hurt in an ankle for example—then it should be conceivable that some sensations lack this extra quality while continuing to possess the other features. Just as we conceive of cold as the converse quality of warmth, could we not also conceive of a converse quality of sensation location such that one might feel pain in an ankle not positively felt to belong to one's own body. If O'Shaughnessy is right, we can make no sense of either possibility.   (Martin, 1995, p. 270)

Likewise, Dokic (2003, p. 325) claims: "The very idea of feeling a pain in a limb which does not seem to be ours is difficult to frame, perhaps unintelligible". However, the problem that we shall now see is that what Martin and Dokic claim as being unintelligible do actually happen.

We shall first consider referred sensations in tools and then turn to sensations that are felt to be located in a so-called "alien" limb.

2.2 The Puzzle of Tools

According to Martin (1993), not only phantom limbs but also tools can be experienced within the apparent confines of one's body. In other words, one can feel sensations to be located in tools in the same way one can feel sensations to be located in phantom limbs. There seems to be no fundamental difference in spatial phenomenology between the two types of sensations. But if Martin does indeed agree with this claim, as he seems to be doing, then he puts his own theory of ownership in a difficult position. His view predicts that the tools should be experienced as being parts of one's own body like phantom hands. Yet there is generally no sense of ownership for tools. We thus have to deny either that we can have referred sensations in tools or that a sense of ownership is possessed by all located sensations.

2.2.1 Referred Sensations in Tools

> The lower animals keep all their limbs at home in their own bodies, but many of man's are loose, and lie about detached, now here and now there, in various parts of the world—some being kept always handy for contingent use, and others being occasionally hundreds of miles away. A machine is merely a supplementary limb; this is the be all and end all of machinery. We do not use our own limbs other than as machines; and a leg is only a much better wooden leg than any one can manufacture. Observe a man digging with a spade; his right forearm has become artificially lengthened, and his hand has become a joint.

> (Butler, 1872, p. 267)

In his utopia *Erewhon*, Butler denies any significant difference between tools and hands.[1] More than a century later, empirical research seems to have confirmed his view (for review, Martel et al., 2016). As it is being picked up, the tool is like any other object: there is no difference between grasping a spade and grasping an apple. However, the control that we have over the spade while using it involves shifting the end-effector

---

[1] By tool, I do not mean any kind of object but only external objects that one actively manipulates (and not simply holds) for a functional purpose (Beck, 1980).

(i.e. what is in interaction with the environment on which one acts) from the hand to the spade itself. The spade is then processed in the same way as one's body parts at the sensorimotor level. This is well illustrated by the following study by Cardinali et al. (2009a). In this study, participants repeatedly use a long mechanical grabber. When subsequently retested while reaching to grasp with their hand alone, the kinematics of their movements is significantly modified, as if their arm were longer than before using the grabber. Moreover, this effect of extension is generalized to other movements, such as pointing on top of objects, although they have never been performed with the grabber. Finally, as Butler well describes it, the arm that is holding the grabber is perceived as longer: after using the tool participants localize touches delivered on their elbow and middle fingertip as if they were farther apart. This result is consistent with another study by Sposito et al. (2012), which has found that after tool use participants mislocalize the centre of their arms as if their arms had increased in size.

A further proof of the embodiment of tools comes from the following simple experiment. Close your eyes and cross your hands over your body midline. As we shall see in more detail in Chapter 4, if your left hand is briefly touched and then your right hand, you take more time and you are less accurate in judging which hand was touched first (Yamamoto and Kitazawa, 2001a). What happens now when one holds two sticks that are crossed with one's hands *uncrossed*, and the two sticks are vibrated one after the other? If the vibration were felt in the hands holding the stick, there should be no conflict (e.g. the vibration on the right hand is on the right), and one should have no difficulty judging which stick was vibrated first. However, this is not what has been found. Participants have difficulties with the sticks crossed (Yamamoto and Kitazawa, 2001b; Yamamoto et al., 2005). This indicates that the vibration is experienced as being located at the tip of the sticks (which are crossed), rather than on the hands that hold them (which are uncrossed). Hence, feeling sensations in tools is not merely a way of speaking (Katz, 1925; Lotze, 1888; Gibson, 1979; Martin, 1993; O'Shaughnessy, 2003; Vesey, 1961; Vignemont, 1917a). As Lotze (1888) vividly describes it:

So, further, does the woodcutter feel, along with the axe's reaction against his hand, its hissing cleaving of the wood; so does the soldier feel his weapon piercing the flesh of his antagonist. (Lotze, 1888, pp. 588–9)

Consider an example hopefully more familiar than piercing the flesh of your enemy: you are cutting carrots into thin slices, but you are having difficulty because your knife is not sharp enough. You then primarily feel the resistance of the carrots annoying you rather than the resistance of the knife in your palm, which is less phenomenologically salient. This is one of many examples of what is known as *referred sensations* in tools. One way to describe them is to say that one feels sensations as of being located in tools, whether it is in the spade, in the axe, in the knife, at the tip of the blind person's white cane, or in prostheses. An amputee, for instance, reported: "I can actually 'feel' some things that come into contact with it, without having to see them" (Murray, 2004, p. 970).

But does the amputee actually feel sensations in his prosthesis in the same way as he feels sensations in his intact hand? One may suggest that the spatial phenomenology of sensations located in tools differs from the spatial phenomenology of sensations located in one's body. Let us ana-lyse the example of the blind man in more detail. One might claim that he only *indirectly* feels at the end of his white cane that there is a bump on the floor in virtue of directly feeling a change of pressure on his palm. It is true that insofar as tactile receptors are on the skin, and not on the tool, referred sensations must involve subpersonal mechanisms of pro-jecting sensations to the tool and recruit different brain processing than non-referred sensations (Limanowski and Blankenburg, 2016). But this is not the same as to say that the blind man only indirectly feels his sensations at the tip of his cane. Compare with the following example. One can *indirectly* hear that the postman is coming on the basis of hearing that the dog is barking and of knowing that the dog barks each time that the postman comes (Dretske, 2006). Perception is then said to be indirect because some of the information about the perceived object or event (e.g. the postman coming) is not embedded in the information about the more proximal object (e.g. the dog barking). By contrast, one *directly* hears that the postman is coming on the basis of hearing his voice. If now we consider the example of the blind man and his white cane, information about the bumps on the floor is embedded in infor-mation about the pressure on the palm. It just needs to be extracted and conceptually structured. No further knowledge is required: there is no need to first categorize the specific pressure that one feels in one's hand, and then infer on the basis of past associations that there is a bump on the floor. Arguably, the first time that the blind man holds his white cane, he can immediately feel the obstacles at the end of his cane. He might not be able to correctly categorize what he feels, but this can be the same when using his own fingers to recognize objects. What is important is that the first time he uses his cane he immediately feels the world in a certain way at the end of his cane (the resistance of the floor, its volume, etc.).

To recapitulate, there is no principled reason why one should conceive of sensations in tools in a way different from the way we conceive of sensations in phantom limbs. Both types of sensations reveal that one can feel sensations beyond the actual limits of the body. There is, however, a major difference. We manipulate tens of tools during the day but do we experience all of them as being parts of our body? It might be that Martin (1993) would reply positively but this seems counter-intuitive. While cooking, we do not have a sense of ownership for the knife and this is so despite our referred sensations in it. Nor does it seem that we feel that we are losing a part of our body when we release it. More generally, we do not experience all the objects that we use as parts of our body.[2] As Botvinick (2004, p. 783) notes wondering about the difference between the rubber hand illusion and tool use:

From this finding [brain activity in multisensory areas], we would predict that tools are represented as belonging to the bodily self. However, the feeling of ownership that we have for our bodies clearly does not extend to, for example, the fork we use at dinner.

Even prostheses are rarely appropriated by amputees. They remain extraneous to them although they can be included within the apparent boundaries of their body (Canzoneri et al., 2013a). Consequently, only 45 per cent of arm amputees choose to use them regularly. Hence, contrary to what Martin claims, not all located sensations possess the sense of ownership: feeling sensations in an object does not suffice for this object to feel as apparently belonging to one's body.

---

[2] I am not saying that it is *impossible* to feel ownership for tools. It may happen after long-term use or under artificial circumstances, as in the following study. Following the same principle of visual capture of touch as the RHI, Cardinali and her colleagues (under revision) induced tactile sensations on the "fingers" of a mechanical grabber after synchronous stroking. As in the RHI, participants reported in the questionnaire that they felt "as if the tool was my hand".

2.2.2 A Special Status for Pain?

There is another difference between referred sensations in tools and phantom sensations. Unfortunately, pain in phantom limbs does exist. But pain in tools does not exist. There are indeed limits to the kind of sensations that one can feel in tools. Obviously you wince when your car bumps into another car but you do not feel pain as being localized in the boot of your car. The fact is that you use tools in harmful situations in which you would not use your limbs. This is so because if a tool is damaged, one does not feel hurt. Hence, the location at which one can feel touch does not always coincide with the location at which one can feel pain. We thus face the following puzzle:

- One can feel touch and pain to be located in a phantom limb;
- One can feel touch to be located in tools;
- And yet one cannot feel pain to be located in tools.

It might then seem that pain is more robustly attached to the body that one experiences as one's own than other types of bodily sensations. As said earlier, amputees experience their phantom limb, in which they can feel pain, as part of their own body, whereas the blind man does not experience the cane, in which he cannot feel pain, as part of his body. One may then suggest that it is only feeling pain in a body part that can be sufficient for experiencing this body part as one's own.

When in pain, I experience disorder as located within a boundary of whose exterior I can have no similar awareness; I am aware that a part of *my* body—not just of *that* body—is disordered. (Bain, 2003, p. 523)

Both Martin and Dokic use the example of pain when trying to show the inconceivability of sensations in a body part that is not felt to belong to one's body. The hypothesis of the privilege of pain can be also found in the psychological literature. For instance, Kammers et al. (2011, p. 1320) claim: "Pain also reminds us of our 'real' body and thus seems to have a special status in body ownership". But what special status?

Painful sensations can grab our attention far more than any other types of bodily experiences, and that by attracting attention to our bodies they can make our sense of bodily ownership more phenomenologically vivid. One may further suggest that there is a sense in which pain is *necessary* for the sense of bodily ownership. As I shall show in Chapter 9, pain endows the boundaries of one's body with a unique significance and if one has never been in pain one experiences one's body as a tool, no longer as something to which one is intimately related. However, this is not the same as to say that feeling pain in a body part is *sufficient* for one to experience this body part as one's own. As we shall now see, one can feel pain to be located in a body part that does not feel to be one's own.[3]

2.3 The Puzzle of Disownership

We saw in Chapter 1 that there is a variety of clinical conditions in which one loses the sense of ownership that one normally experiences for one's own body. The critical question is why this happens. One possible answer is that the individuals no longer feel sensations in their limbs, and thus no longer experience these body parts as their own. This reply may explain the feeling that we can have after falling asleep on our arm. However, it predicts that patients who no longer experience their limbs as their own should feel nothing there and this is not always the case: some patients can actually feel sensations to be located in their so-called "alien" limbs.

Let us analyse in more detail the case of patients suffering from somatoparaphrenia. Many can still feel pain and even cry out if the examiner pinches their "alien" hand (Melzack, 1990). For instance, one patient asked his doctor:

P: I still have the acute pain where the prosthesis is. E: Which prosthesis? P: Don't you see? This thing here (indicating his left arm). The doctors have attached this tool to my body in order to help me to move [...] Once home could I ask my wife, from time to time, to remove this left arm and put it in the cupboard for a few hours in order to have some relief from pain? (Maravita, 2008, p. 102).

There is no doubt here that the patient was experiencing pain and that he was locating his pain in his left arm, and yet there is also little doubt that he was not experiencing his left arm as his own. Touch is more frequently affected in somatoparaphrenia, but it can also be preserved in some patients. For instance, Bottini et al. (2002) described the case of a

---

[3] I am not referring here to situations of empathy or emotional contagion. It is true that if I see the door of the car closing on your foot, I will most probably shrink. Nonetheless, I do not feel pain as being located in your foot; I only experience *vicarious* pain, which is isomorphic to your pain, but to some extent only (Vignemont and Jacob, 2012). For more detail, see Chapter 7.

somatoparaphrenic patient, F. B., who denied ownership of her left hand and attributed to her niece. When told she was going to be touched on her left hand, F.B. systematically failed to detect the touch. However, if the examiner told her that she was going to be touched on her niece's hand and instead her left hand was touched, F. B. reported feeling the touch with a degree of accuracy much greater than chance. After the experiment, when asked how she could feel touch in someone else's hand, she replied, "Yes, I know, it is strange" (p. 251). Another somato-paraphrenic patient was able to detect light tactile stimuli on his "alien" hand and even to distinguish between "sharp" and "dull" stimuli (Cogliano et al., 2012). Interestingly, this patient reported feeling 30 per cent of the stimuli applied to his "alien" hand on his contralateral hand, which he experienced as his own. The spatial transposition might be a way to make sense for him of the fact that he could feel sensations on a body part he did not experience as his own. Still, for the remaining 70 per cent of the tactile stimuli he localized them on his "alien" hand.

Hence, contrary to Martin's prediction, patients with somatoparaph-renia can report feeling sensations—including pain—to be located in the hand that they disown. Bodily experiences lack the first-person at one level: patients are aware that *they* feel bodily sensations, but they are not aware that they feel them in *their* body. Their bodily experiences thus fail to ground self-ascriptive bodily judgements. Feeling sensations to be located in a body part does not suffice to account for the first-person character of the sense of bodily ownership.

One may then try to save Martin's view by adding a supplementary condition for the sense of ownership, namely attention (Kinsbourne, 1995; Hochstetter, 2016). In Chapter 1 we saw that thanks to vestibular stimulation patients can temporarily regain their sense of ownership for their hand. Interestingly, vestibular stimulation is known to manipulate space-based attention. One may then suggest that what somatoparaph-renic patients lack is not the feeling of ownership for their own hand but more simply their ability to pay attention to their hand. In favour of this attentional hypothesis is the frequent association between somatopar-aphrenia and a related condition named personal neglect, in which patients fail to pay attention to the left side of their body. However, not all somatoparaphrenic patients suffer from attentional disorder (Vallar and Ronchi, 2009). Furthermore, it has been shown that bodily attention does not suffice to regain the sense of bodily ownership. Instead of using vestibular stimulation, whose mechanisms are still partly mysterious, Moro et al. (2004) simply positioned the patients' neglected "alien" left hands in their non-neglected right hemispace. They found that the patients were able to report with perfect accuracy when they were touched on their left "alien" hand but despite paying attention to their hand and feeling that they were touched there, they still failed to feel it as being their own. Hence, there is more to the sense of bodily ownership than attention to one's body.

Alternatively, one may attempt to save Martin's view by interpreting the denial of ownership in somatoparaphrenia as being *irrational*. One may, for instance, propose that somatoparaphrenic patients do experience their hand as their own, but merely overlook their experience of ownership because of reasoning deficits, thus leading them to have delusions of disownership. But can reasoning deficits suffice to explain their delusions? As mentioned in Chapter 1, one needs to distinguish between the factors that trigger the initial implausible thought (and thus contribute to explaining the thematic content of a particular delusion), and the factors that explain the uncritical adoption of the implausible thought as a delusional belief (Langdon and Coltheart, 2000; Coltheart et al., 2011). Abnormal rationality can only account for the feeling of confidence in the delusional beliefs, but not for their content. In addition, one should not forget that some of the disorders that I described occur only at the phenomenological level. In other words, not all patients with disownership are delusional. For example, Ian Waterman, the deafferented patient, did not believe that his body was alien; he only felt *as if* it did not belong to him. He suffered only from a peripheral disorder and he had no psychiatric disorder or brain lesion. It is hard to see how one could explain his sense of disownership in terms of abnormal rationality. And yet he had some spared bodily sensations, including thermal sensations and pain, which did not suffice for eliciting a sense of bodily ownership.

Ian has described how he would sometimes wake to feel a hand on his face and not know to whom it belongs. Until he realised it was his own, the experience was momentarily terrifying. Since he has normal perception [. . .] of warmth in the hand, it is interesting that he cannot, or does not, use warmth of the hand alone to identify self from non-self.   (Cole, 1995, p. 85)

One may then be tempted to argue that the felt location of his preserved sensations is abnormal: he has no proprioception left and thus, although he can feel warmth in his hand, he cannot localize the warm hand within an external frame of reference without vision. Because of this deficit, his spared bodily sensations would not be able to ground a sense of bodily ownership.[4] The problem with this reply, however, is that his body felt as alien only during the first few months following his neuropathy, but not afterwards. Had he regained proprioception? The answer is negative; only his control over his body had changed: he had learned to exploit visual information to compensate the lack of proprioception in order to plan and guide his bodily movements. The origin of his lack of ownership at the beginning must thus be found elsewhere than in his sensory phenomenology. Furthermore, other patients with preserved bodily sensations can also experience a sense of disownership, including patients with xenomelia, who have neither reasoning deficit nor any sensory deficit.

To recapitulate, Martin's argument was the following:

- If "the sense of ownership is a positive quality over and above the felt quality of the sensation and the location".
- Then it should be conceivable that "one might feel pain in an ankle not positively felt to belong to one's own body".
- This is inconceivable.
- Hence, there is no sense of ownership over and above the felt location of bodily sensations.

However, we have just seen that although bodily experiences and the sense of ownership normally go together, they can sometimes be dissociated: feeling bodily sensations, even painful sensations, in a part of one's body or in an object does not systematically suffice for the sense of ownership. It thus seems that the sense of ownership is over and above the felt location of bodily sensations.

---

[4] Along the same lines, it has been suggested that the denial of ownership in somatoparaphrenia is specifically linked to the disruption of the first-person perspective (Salvato et al., 2017). In favour of this interpretation, it has been found that when patients see their "alien" hand in a mirror, thus from a third-person perspective, they temporarily recognize their hand as their own (Fotopoulou et al., 2011; Jenkinson et al., 2013). However, one may doubt that somatoparaphrenic patients *feel* ownership for the hand that they see in the mirror. Visual experiences of one's body do not present one's body as one's own the way bodily experiences do (Martin, 1992; Brewer, 1995). They may judge that this is their own body but they do not experience it.

2.4 A Feeling of Presence

Let us go back to Martin's proposal:

For awareness of one's body as one's body involves a sense of its being a bounded object within a larger space, and that just is to locate it within a space of tactual objects. (Martin, 1992, p. 213)

I have argued that one does not have to accept Martin's gloss of the awareness of bodily boundaries in terms of ownership. So what dimension of bodily awareness, if not ownership, does Martin describe? He claims that by being aware of the locations at which one can feel sensations and those at which one cannot, one becomes aware of the boundaries of one's body but also of a space larger than one's body, a space within which this body is located along with other objects. In this sense, one's body is perceived as being one object among others. I propose that this anchoring in three-dimensional external space corresponds to a feeling of bodily presence instead of a feeling of bodily ownership.

The notion of *feeling of presence* has originally been proposed to characterize the distinctive visual phenomenology associated with actual scenes and objects, which is lacking in visual experiences of depicted scenes and objects (Noë, 2005; Matthen, 2005; Dokic, 2010). Noë (2005) notes that when facing a tree, one sees only one side of it and yet one is aware of the presence of the whole tree. On his view, to be aware of the presence of the tree is then to know that if one walks around the tree, one can see the other side. By contrast, when facing the painting of a tree, there is no hidden side of the tree that one could access. Thus, one does not experience the depicted tree as being present. Seeing an object as present involves being aware of it as a whole object located in three-dimensional space, as an object that one can explore from different perspectives and that one can grasp, while seeing a picture of the same object only involves being aware of its material surface with certain configurational properties.

In the same way that there is a feeling of presence associated with visual experiences of actual scenes and objects, I suggest that there is a feeling of *bodily presence* given by the spatial phenomenology of bodily experiences. For instance, when something brushes our knee, not only do we feel a tactile sensation, we also become suddenly aware of the presence of our knee as being located in egocentric space, as a body part that we can reach and grasp. The existence of such a feeling is well illustrated by amputees who still feel as if their lost limb were still there, physically present:

> To anyone looking at me, I have no arm. But I can feel the entirety of my phantom hand and arm.   (Mezue and Makin, 2017, p. 34)

The reverse may also happen in depersonalization:

> I do not feel I have a body. When I look down I see my legs and body but it feels as if it were not there.   (Dugas and Moutier, 1911, p. 28 in Billon (2017))

Interestingly, depersonalized patients often feel the urge to touch their body or to pour hot water on it to reassure themselves of its existence: "As I sense it I have the need to make sure and I rub, touch, and hurt myself to feel something" (Sierra, 2009, p. 29). Self-touch is used as a means to try to re-establish their feeling of bodily presence. In the same way, we might pinch ourselves to make sure that we are not dreaming. Now one may note that in depersonalization the lack of feeling of bodily presence is associated with a lack of feeling of bodily ownership. It may then be easy to conflate the two types of feelings. It is actually difficult to conceive of a situation in which one could feel as one's own a body that one does not feel as being present. This, however, does not show that they are one and the same feeling.

Let us return to the cases that we described in this chapter. I have argued that one can experience referred sensations in tools. Thanks to these sensations one experiences the tools as being part of a larger space on which they can act. One thus experiences them as being "here", present in three-dimensional external space. Yet one may feel no ownership for them. Consider now somatoparaphrenia. Because of their preserved sensations patients with somatoparaphrenia can feel the "alien" limb as being intrusive, as being always here. A patient, for example, complained:

> It was very difficult to begin with . . . to live with a foot that isn't yours . . . It's always there, always present . . .   (Halligan et al., 1995, p. 176)

One may then suggest that the feeling of bodily presence is preserved but not their feeling of ownership. There seems to be a priority of presence over ownership: one first needs to feel one's body as being here before feeling it as being one's own. It is therefore possible to experience presence without ownership but not the reverse. The awareness of bodily boundaries, which is at the origin of the feeling of bodily presence, is thus a necessary condition for the sense of ownership but it is not a sufficient one.

What these dissociations reveal is the necessity to distinguish what Peacocke (2014) calls the degree 0 of nonconceptual self-representation, which can be described in terms of "this body", and the degree 1, which involves *de se* content of the type "my body". They further show that it does not suffice to experience the spatial phenomenology of bodily sensations, and thus to have a feeling of presence, for having states with degree 1 content:

> The content *this leg is bent*, even based on proprioception, or capacities for action with the leg, or both, is not yet the content *my leg is bent* [. . .] So the question becomes pressing: What more is required to make a nonconceptual content *c* the first-person nonconceptual content *i*? This is equivalent to the question: What is it for an organism to be at Degree 1 of self-representation rather than Degree 0? It is also equivalent to the question: What minimally brings a subject into the (referential) content of a mental state?   (Peacocke, 2017, p. 292)

In other words, the awareness of one's body as one's own does not simply consist in the awareness of one's body as a bounded object. What somatoparaphrenia and the other borderline cases that I described reveal is that we should not confuse spatial awareness and self-awareness, presence and ownership: one can be aware of the boundaries of the body without being aware of the boundaries of the body *qua* one's own (Dokic, 2003; Serrahima, forthcoming):

> For, we could still ask: on what grounds does the subject take the body at one of the sides of the boundary—the body instantiating "Pressure" in the worktable example—to be her own in the relevant sense, while taking the object at the other side—instantiating "Roughness"—to not be hers? Why should any side of the perceived boundary have the special import it has?   (Serrahima, forthcoming)

In other words, what makes one side of the boundaries, the side in which I feel sensations, mine? Or, as Peacocke (2017) puts it, "What more is required?" In Part III I shall reply to this question by appealing to the evolutionary significance of bodily boundaries for the organism.

2.5 Conclusion

At first sight, Martin provided a promising reductionist account of the sense of bodily ownership by explaining the first-personal character of bodily ownership in spatial terms. However, I have highlighted in this chapter two puzzles that it fails to account for. The puzzle of tools first: for what reason does one generally experience no ownership towards tools although one can feel some sensations as being located in them? The puzzle of disownership secondly: how can one experience disowner-ship towards one's own limbs despite feeling sensations in them? What are then the alternatives to Martin's view? They can go in two directions: either towards a more conservative conception or towards a more liberal one.

Consider first the conservative option. Martin defends a *constitutive* relationship between the phenomenology of ownership and the spatial phenomenology of bodily experiences but one may propose that the relationship is simply *causal*. On this alternative view, the sense of ownership is normally accompanied by the awareness of other bodily features, which can then be taken as signs that this is one's own body. The location of bodily sensations may then be merely one sign among others on the basis of which one can report ownership but none of these signs is conceived of as being necessary to the sense of ownership. Unlike Martin's view, this weaker conception almost collapses into eliminati-vism. Martin indeed reduces the phenomenology of ownership but he does not eliminate it, while in the causal account there is no such thing as a phenomenology of ownership, there is only a bundle of bodily properties that provide causal or epistemic grounds (but not constitutive ones) for the sense of ownership. This view is less vulnerable than reductionist accounts because no dissociation can be taken as counter-evidence. The problem, however, is that it makes so few theoretical commitments that it is actually hard to conceive of what could falsify it. It also becomes even more difficult to understand the sense of disownership reported by some patients: if nothing is necessary to the sense of ownership, how can one lose it?

The alternative is then to go in the opposite direction and defend a conception more liberal than Martin's. Martin indeed defends the view that the phenomenology of ownership is not an additional quality to the sensory qualities of bodily experiences and that it is "somehow already inherent within them" (Martin, 1995, p. 278). We have seen, however, that the quality of ownership cannot be reduced to a sensory quality. What we then have to do is to go beyond the sensory dimension of

Does it feel different when one is aware that this body is one's own and when one is not?

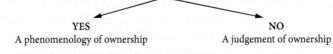

YES   NO
A phenomenology of ownership   A judgement of ownership

Is it an inherent sensory quality of bodily sensations?

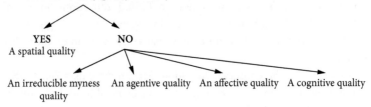

YES   NO
A spatial quality

An irreducible myness   An agentive quality   An affective quality   A cognitive quality
quality

Figure 2.1. A brief overview of the debate.

bodily awareness and systematically explore its agentive and affective dimensions (see Figure 2.1). This shall be my task in this book. But first I shall consider some of the epistemological peculiarities of the sense of ownership, and more particularly, their immunity to error through misidentification relative to the first-person.

3

The Immunity of the Sense of Ownership

Consider the following case. I judge that I am anxious. Can I be mistaken? The kind of errors that I can make depends on the grounds of my judgement. It may be that I do not especially feel anxious but my psychoanalyst told me that I am. However, he is not a reliable source of information about myself and he has actually confused me with another of his patients, who is the person suffering from anxiety. Consequently, I have incorrectly self-attributed to myself the property of being anxious. Alternatively, it may be that I judge that I am anxious because I feel that way. My judgement is not infallible: I can miscategorize my emotional state and be actually excited rather than anxious. However, there is one mistake that I cannot make: I cannot be wrong about the person that is experiencing this specific emotional state. Since I am aware of my emotion through introspection, it cannot be anyone else than me who feels it. This specific type of infallibility is known as *immunity to error through misidentification relative to the first-person* (hereafter IEM) (Shoemaker, 1968).

The notion of IEM has been developed primarily to account for the epistemic behaviour of self-ascription of mental states, but let us see whether it can be applied to the self-ascription of bodily states.[1] I form the judgement that my legs are crossed while I am seated around a small table with several other persons. Can I be mistaken? If I make this judgement on the basis only of my seeing legs that are crossed that I take to be mine, then the answer is positive. I may be confusing my legs with the legs of the person seated next to me. By contrast, if I make this judgement

[1] Here I mean the ascription of physical bodily properties (e.g. body size, weight, posture, etc.), and not of bodily sensations.

because I feel my legs as being crossed, then the answer is negative. I cannot be wrong about whose legs are crossed when I feel them that way. Since I know that they are crossed through proprioception, they cannot belong to anyone else than me. My bodily judgement is then immune to error through misidentification relative to the first-person (Bermúdez, 1998; Brewer, 1995; Cassam, 1995; Evans, 1982; Vignemont, 2012).

Although the epistemic difference between vision and proprioception is intuitive, one may question its systematicity. On the one hand, we have seen in Chapter 2 that the judgements of ownership that are grounded on bodily sensations can be erroneous. On the other hand, one might wonder whether the body cannot be visually presented in such a way that it can be only one's own body. In this chapter I will reconsider which experiences can ground bodily judgements that manifest IEM. This will help me analyse the relationship between the phenomenological phenomenon of bodily ownership and the epistemic phenomenon of IEM. I will argue that it is important to keep the two phenomena apart: one should not conceive of the experience of ownership as the phenomenological counterpart of bodily IEM.

First, however, I need to clarify the notion of IEM that I will use here. Unfortunately, there is little agreement on its interpretation. One may analyse it in terms of conceptual truth about mental concepts, in terms of rational structure of self-knowledge, and/or in terms of reliability of causal grounds. Here I shall assume that a self-ascriptive judgement *I am F* is immune to error through misidentification relative to the first-person only if it is appropriately grounded, that is, if in ordinary circumstances one cannot have rational doubts about the person who instantiates the property F given the way that one has gained information about F. There are stronger readings of IEM, and one might argue that the kind of IEM that bodily self-ascriptions display is weak compared to the IEM displayed by other types of self-ascriptions (see O'Brien, 2007 for discussion), but this is the one that I shall use from now on.

3.1 Bodily Immunity from the Inside

We have what might be described as a general capacity to perceive our own bodies, although this can be broken into several distinguishable capacities: our proprioceptive sense, our sense of balance, of heat and cold, and of pressure. Each of these modes of perception appears to give rise to judgments which are immune to error through misidentification. [...] There just does not appear to be a gap between the subject's having information (or appearing to have information), in the appropriate way, that the property of being F is instantiated, and his having information (or appearing to have information) that he is F.

(Evans, 1982, pp. 220–1)

Bodily senses guarantee IEM because they provide a privileged informational access to one's own body only. This is not to say that they do not carry information about anything else. The sense of balance is about the relation between the external world and one's body, and touch is about the relation between the touching object (*touchant*) and the touched body part (*touché*). However, the subject's body is always part of the relation. Even if I can have a proprioceptive experience of the location of another individual's hand simply by holding it with my eyes closed, my experience is only indirect and derives from having a direct proprioceptive experience of the location of my own hand. Because of this privileged relation, the judgement "my legs are crossed" is not grounded in the judgement "these legs are crossed" and in the identification "these legs are mine". Proprioceptive experiences suffice to justify bodily self-ascriptions such that no intermediary process of self-identification is required. I shall call this view the bodily IEM thesis.

One may challenge the bodily IEM thesis by conceiving a case of cross-wiring such that A is connected through A's proprioceptive system both to A's body and to B's body. Then A can feel B's legs being crossed. In this scenario, knowing that the legs are crossed via proprioception (or quasi-proprioception) no longer guarantees for A that those are A's legs that are crossed. This thought experiment shows that bodily IEM can be only de facto, true in worlds in which bodies are connected only to one subject. Bodily IEM ultimately derives from the basic biological fact that one is connected only to one's body through proprioception. This is why it is rational for the subject to self-ascribe bodily properties when she has access to them through proprioception. There are, however, other cases that might be conceived of as a breach of the bodily IEM thesis, even in its de facto reading. The first type consists in *false negative* errors: one does not self-ascribe properties that are instantiated by one's own body. This seems to be typically the case in patients with somatoparaphrenia, who fail to self-ascribe the hand that they feel being touched or in pain. The second type consists in *false positive* errors: one self-ascribes properties that are instantiated by another's body. This is possibly the case in the RHI, when participants self-ascribe properties instantiated by a rubber hand (although it is only at the experiential level). This is also possibly the case in what is known as the embodiment delusion, in which patients judge that another person's hand is their own. However, I will argue that none of these errors undermines the bodily IEM thesis.

3.1.1 False Negative: Somatoparaphrenia

In Chapter 2, I described the patient FB who suffers from somatoparaphrenia. She can feel tactile sensations when her left hand is touched but she judges that it is her niece's hand that is touched (Bottini et al., 2002). In other words, she has the appropriate grounds that entitle her to self-ascribe the hand that is touched, and yet she does not judge that the hand that is touched is part of her own body. It is tempting to think that her error is exactly of the type that should not happen if the bodily IEM thesis were true. Let us thus analyse her case in more detail.

One way to interpret her error is in terms of Pryor's notion of error through *which-misidentification* (Pryor, 1999). There is which-misidentification if one is justified in making an existential generalization such that there is an *a* that is F, although one is wrong in figuring out which object instantiates the property F. Here, the tactile sensation experienced on the hand justifies FB in making an existential generalization such that there is *a hand* that is touched. However, she is wrong about which object instantiates the bodily property because she believes that it is her niece's hand. But does this show that judgements of bodily ownership are sensitive to error through which-misidentification, even when appropriately grounded? No, because somatoparaphrenia is actually irrelevant for the discussion of bodily IEM.

The patient has the appropriate grounds that guarantee the IEM of bodily judgements (e.g. she experiences tactile sensations). Yet the presence of these appropriate grounds does not suffice to guarantee that she will use them and will make the right bodily judgement. FB may indeed either ignore them or have other reasons that she takes to defeat them. Roughly speaking, it is not merely because one has plenty of good reasons to believe that smoking causes lung cancer that one actually believes it. And it is not because one does not believe in the harm caused by tobacco that one does not have good reasons for believing it. The bodily IEM thesis claims that if the judgement derives from the right grounds, then it is immune to error. It does not make any claim about whether one actually makes the judgement or not. As such, one may doubt that it makes any claim about false negative errors, such as somatoparaphrenia.[2] What it does target are false positive errors, which we shall now consider.

3.1.2 False Positive: the RHI

In the RHI, one self-attributes a hand that does not belong to one. Interestingly, it has been argued that this type of bodily illusion shows that proprioceptive-based bodily self-ascriptions depend on the identification of the body:

Thus it demonstrates the case of judgement based on somatic proprioception, which is mistaken because of misidentification, and therefore constitutes a counterexample to the immunity thesis.   (Mizumoto and Ishikawa, 2005, p. 12)

Before going into the detail of what is happening in the RHI and seeing whether it warrants this conclusion, let us note that it is questionable whether participants actually self-attribute the rubber hand. As argued in Chapter 1, they feel as if the rubber hand were part of their own body, but they do not believe it. IEM does not preclude any kind of errors, only errors relative to the first-person, and it is not clear here that there is such an error at the doxastic level. Still, one subject in the original study spontaneously reported: "I found myself looking at the dummy hand *thinking it was actually my own*" (Botvinick and Cohen, 1998, p. 756, my emphasis). So let us grant that they sometimes self-attribute the rubber hand at the doxastic level.[3]

---

[2] Interestingly, the alleged counterexamples that have been proposed against IEM to date have always been false negatives, such as thought insertion, anarchic hand sign, and delusion of control (Campbell, 1999; Marcel, 2003). On the basis of these, some claim that the sense of agency is not immune to error (Marcel, 2003; Jeannerod and Pacherie, 2004), while others claim that those cases are of no threat to IEM (Peacocke, 2003; O'Brien, 2003).

[3] Another issue is that the setting of the RHI is such that it can be conceived of as a kind of deviant causal chain, which, according to Evans (1982, pp. 184–8), cannot be taken as evidence against the IEM thesis. Evans describes the following example: he is wearing earphones that allow him to hear noise in a different room. His judgement "here it is noisy" then seems to result from an identity assumption ("location x is noisy" and "location x is here"). If so, the indexical thought cannot be immune to error. But according to Evans, such an example cannot be taken as evidence against the IEM thesis of the indexical "here" because of the abnormal deviant causal chain of auditory perception in this case. The IEM thesis rests on the assumption of the reliable working of our perceptual systems under *normal conditions*.

Still Mizumoto's and Ishikawa's conclusion is problematic. In order to best point out the difficulties that it faces, I will reconstruct the RHI using Pryor's notion of *de re* misidentification (Pryor, 1999). There is *de re* misidentification if one makes a false identification assumption about two particular objects such that y = x. One can then propose the following analysis of the RHI:

- *Singular proposition*: the paintbrush P strokes the rubber hand;
- *Identity assumption*: my hand = the rubber hand;
- *Bodily self-ascription*: the paintbrush P strokes my hand;
- But actually, my hand ≠ rubber hand.

The participants are right to believe that the rubber hand is in contact with the paintbrush P, but they are wrong to identify the rubber hand with their own hand. However, even if the analysis of the RHI in terms of *de re* misidentification were right, this would not show a breach of proprioception-based IEM because the singular proposition is not based on proprioception. Indeed, the RHI involves *visual capture* of touch and proprioception. It is vision, rather than proprioception or touch, that provides the information that the property is instantiated (i.e. the rubber hand is stroked by the paintbrush *P*). Bodily senses play a role, but only at the level of the identity assumption, when they are compared to vision (i.e. temporal synchrony and spatial congruence between the visual experience and the tactile experience). Hence, if one assumes that vision is a way of gaining bodily information that does not guarantee bodily IEM, it is possible for errors through *de re* misidentification to happen.

There is, however, a non-visual version of the RHI (Ehrsson et al., 2005). In the somatic RHI participants are blindfolded. The experimenter strokes the participants' biological right hand and moves their left index finger so that it strokes the rubber hand in synchrony. Participants report feeling that they are touching their own right hand. When asked to point to their right index finger after the illusion, the blindfolded participants mislocate their right hand in the direction of the location of the rubber hand. At first sight, the somatic RHI seems to be more challenging for the bodily IEM thesis. It is not based on vision but on touch. However, touch carries information both about the external world and about the body. In the case of the somatic RHI, participants have to rate their sense of ownership for the hand that they stroke and feel under their fingertips (i.e. the touched hand). They are not asked about the hand that is actively performing the stroking movement (i.e. the touching hand). The information that they have about the rubber hand thus exclusively relies on the exteroceptive content of touch, which does not guarantee bodily IEM. To show this, imagine that only your left hand is anaesthetized. While you are in the dark, your right hand touches a hand. Whose hand is that? You can be mistaken and judge that it is your own left anaesthetized hand, although it is someone else's hand. Therefore, the somatic RHI is no more a counterexample for the bodily IEM thesis than the visual RHI is.

One can also wonder at which stage the identity assumption intervenes. More precisely, one can contrast two scenarios. On the one hand, bodily self-ascription follows from the singular proposition and the identity assumption. On the other hand, the identity assumption follows from bodily self-ascription. It then seems that it is only the former scenario that might constitute a counterexample to the bodily IEM thesis. Only then indeed does bodily self-ascription include an identification component. But the RHI in fact corresponds to the latter scenario. In other words, I do not feel touch on the rubber hand because I feel that the rubber hand is my hand. I feel that the rubber hand is my hand because I feel touch on it. Let me briefly explain why. As I will argue in more detail in Chapter 6, the RHI results from multisensory integration, which does not rely on self-identification. It seems unlikely that the participants must feel that their hand is F, see that the rubber hand is F, erroneously judge that their rubber hand is their own hand, and integrate what they feel with what they see. The experimental data suggest that the similarity between the rubber hand and the participants' own hand can be minimal. Most visual dissimilarities, including different hand shape, laterality, and skin complexion, at most reduce the illusion but do not prevent it (Longo et al., 2009a; Petkova and Ehrsson, 2009). The identity assumption is not a prerequisite of the visuo-somatosensory integration that leads to bodily self-ascription; it is a consequence of it.

3.1.3 False Positive: the Embodiment Delusion

To recapitulate, one is clearly mistaken in self-attributing the rubber hand but the grounds of the judgement are not exclusively internal. Hence, the RHI does not challenge the hypothesis that the sense of ownership is immune to error. There is, however, another case of false positive, which might put more pressure on the bodily IEM thesis. Some somatoparaphrenic patients can self-attribute another individual's hand, although they fail to self-attribute their own hand. This may occur spontaneously with the doctor's hand, for instance, or this may be induced experimentally by using the RHI (e.g. Bolognini et al., 2014; d'Imperio et al., 2017; Fotopoulou et al., 2011). It has also been recently found that other patients, who do not explicitly deny ownership of their own hand, can report feeling another person's left arm as their own under certain circumstances (Garbarini et al., 2017).[4] Let us call the other person's left hand the "embodied" hand.

> E: (The examiner put her left hand under the patient's one) Try to touch your left hand with your right one, please. P: Here it is! (He caught the examiner's hand). E: Whose hand are you touching? P: It is mine.   (Cogliano et al., 2012, p. 765)

When the "embodied" hand moves, patients report feeling their hand moving. When asked to reach for their left hand, they reach for the "embodied" hand. When asked to name the colour of the object in front of their left hand, they name the colour of the object in front of the "embodied" hand (Garbarini et al., 2012).

These patients can also report feeling sensations as located in the "embodied" hand. In one study, noxious stimulation was applied both to the patients' hand and to another person's hand. It has been found that when the patients experienced the other's hand as being their own, they reported feeling pain to the same intensity as when their own hand was hurt and when the "embodied" hand was hurt (Pia et al., 2013). Their subjective report was correlated with their physiological reaction (Garbarini et al., 2014).

Hence, one might claim that the patients felt pain in a body part that did not belong to their body. Does the embodiment delusion constitute a counterexample to the bodily IEM thesis? No more than the RHI. Even if patients can report feeling pain in another person's hand, this is only if they see it on the table next to their own left hand. If they do not see it or

---

[4] The embodiment delusion occurs only when the patients see the other person's hand in front of them next to their own left arm. Then only do they deny ownership of their own hand. But when the other's hand is located elsewhere, the patients are no longer delusional: they neither attribute the other's hand to themselves nor deny their hand to be their own.

if they see it at another location, they no longer feel that the hand belongs to them or that it hurts. Once again, the singular proposition is based on vision, instead of nociceptive information, and vision is not an appropriate ground for judgements about pain.

To summarize, neither somatoparaphrenia nor the RHI nor the embodiment delusion qualify as counterexamples to the bodily IEM thesis. Bodily experiences guarantee the IEM of the sense of ownership. What the bodily IEM thesis does not predict, however, is that there are other types of experiences, and more specifically visual experiences, that can also guarantee bodily IEM.

3.2 Bodily Immunity from the Outside

Ernst Mach famously reported stepping onto a bus and noticing a man who looked like a shabby pedagogue. In fact, he was seeing himself in a large mirror at the far end of the bus. We can indeed easily fail to visually recognize our body. Conversely, we can mistakenly believe that the body that we see is our own. For instance, if we see a broken arm, we can mistake it for our own, although it is another person's arm that is mixed up with ours. In this case, our visual experience does not entitle us to believe that our own arm is broken (Wittgenstein, 1958). On the basis of such examples, it is classically assumed that no judgements based on external senses are immune to error through misidentification, but I will now describe some exceptions to this rule.

3.2.1 Self-Location

The psychologist Gibson (1979) was a staunch advocate of the dichotomy between external senses and bodily senses, arguing that even while looking at the external world, our visual experiences carry information about our own body:

I maintain that all the perceptual systems are propriosensitive as well as exterosensitive, for they all provide information in their various ways about the observer's activities... Information that is specific to the self is picked up as such, no matter what sensory nerve is delivering impulses to the brain.

(Gibson, 1979, p. 115)[5]

---

[5] By propriosensitive, Gibson does not refer to proprioception as we have defined it, but more generally to sensitivity to information about one's body.

Gibson appeals to phenomena such as visual kinesthesis in order to show that thanks to self-specific invariants in the optical flow of visual information (e.g. rapid expansion of the entire optic array), we can see whether we are moving, even though we do not directly see our body moving. In the so-called "moving-room" experiments, participants are placed in rooms whose walls and ceilings can be made to glide over a solid and immovable floor. Participants cannot see their feet or the floor. When they see the walls moving backward and forward, they experience themselves as moving back and forth. Based on their visual kinesthesis, the participants may doubt whether they are moving or not. But they cannot doubt whether it is their own body that they experience as moving back and forth. Arguably, visual kinesthesis grounds judgements about one's bodily movements that are immune to error through mis-identification (Bermúdez, 1998).

Visual experiences of the environment can guarantee the IEM of a second class of self-locating judgements when they represent the environment within an *egocentric frame of reference*. The egocentric perspective carries self-specific information about the location of the perceiver. Hence, when I have the visual experience of a tree in front of me, I am entitled to judge that I am standing in front of a tree. I can be wrong that it is a tree (it can be the painting of a tree, for instance). But, as noted by Evans (1982, p. 222), I cannot have the following doubt: "someone is standing in front of a tree, but is it I?". Evans concludes that visual experiences of the external world from an egocentric perspective guarantee the IEM of self-locating judgements (see also Cassam, 1995).

3.2.2 Seeing One's Body

That visual experiences of the *environment* can ground bodily IEM is one thing, but that visual experiences of the *body* can ground bodily IEM is another. Through vision, one generally has access to one's body as well as to other people's bodies so that there is a gap between visually knowing that a body is F and visually knowing that it is my own body that is F. Most examples in the literature appeal to mirrors to disqualify vision (Mach in the bus, for instance), but one may wonder whether visual experiences of the body might be able to guarantee IEM in more eco-logically valid situations. Here it is helpful to go back to the explanation of bodily IEM of proprioceptive judgements. As said earlier, proprioception guarantees only de facto bodily IEM because of some biological facts about the human body (i.e. the proprioceptive neural system is connected only to one's own body). In the same vein, one may suggest that there are some other basic biological facts that secure the link between certain visual contents and one's own body.

Let us draw a parallel with self-portraits. Most self-portraits are what the artist looks like from several feet—she looks in a mirror and draws what she sees there (e.g. Van Gogh's self-portraits). There is, however, another kind of self-portrait. Ernst Mach drew himself without using a mirror—he drew what he looked like from his own point of view, from zero distance. His self-portrait is faceless, as if the universe were growing out of him. He represents the point of view that he has on his body when looking down to it, a point of view that is well described by Frege (1956, pp. 304–5):

It seems to me as if I were lying in a deck-chair, as if I could see the toes of a pair of waxed boots, the front part of a pair of trousers, a waistcoat, buttons, part of a jacket, in particular sleeves, two hands, the hair of a beard, the blurred outline of a nose.

This specific visuo-spatial perspective is such that it is anatomically impossible that it could be another individual's body that Frege sees in ordinary circumstances (in the absence of deviant causal chains). The combination of the angle and the distance from which his body is visually represented is self-specific. Visual experiences with such a perspective can thus guarantee bodily IEM. Frege cannot rationally doubt that this is his own nose when he sees it in such a way.[6]

These examples indicate that vision can guarantee bodily IEM if one has the right visuo-spatial perspective on the seen body (see Figure 3.1). The right perspective has to be first-personal. One must see from one's own point of view (e.g. toes pointing forward), and not from other people's point of view (e.g. toes pointing towards oneself). This explains why the way one sees one's body in the mirror, which is from a third-person perspective, is sensitive to error through misidentification. However, we can visually represent many objects, including many bodies, from a first-person point of view. One must then distinguish two types of visual experiences from a first-person perspective: neutral and self-specific. Visual experiences are

---

[6] Evans (1982) seems to be sympathetic to this view. He briefly alludes to it in a footnote and includes in the list of appropriate grounds for bodily IEM the movement of "looking down to one's body" (p. 220 fn 26). See also Peacocke (2012).

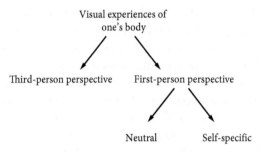

Figure 3.1. Seeing one's body.

neutral when one sees one's body at a location and with an angle that are anatomically compatible both with seeing one's body and with seeing other bodies. For instance, when I see a hand in front of my eyes, it can be either my own hand or the hand of another individual who is standing just behind me. Visual experiences are self-specific when one sees one's body at a location and with an angle that are anatomically compatible only with one's own body under ordinary circumstances (without mirrors, virtual reality device, or other tricks). For instance, when I see my nose while closing one eye, it can be only my own nose. It is only in that latter case, when they are self-specific, that visual experiences guarantee bodily IEM.

Hence, one should not rule out all visual experiences as being inappropriate grounds for bodily IEM. There are some visuo-spatial perspectives from which we can see our body such that we cannot doubt that the body that we see is our own. Indeed, the body that we see that way has always been our own. Our perceptual system has been calibrated by this invariant fact, making the process of self-identification superfluous. The invariance of the relation is of major importance here. Let us imagine another world in which its inhabitants can see their nose as well as other people's noses when they close one eye from the same specific angle. Then their visual system would have been calibrated by the fact that their nose is located in such a way that it could be another person's nose. Their visual experiences would then not guarantee bodily IEM. And if suddenly they could see only their own nose that way, they would still go through a process of identification, and possibly misidentify it, because that is what their visual system has always been doing. We are not like that. Our visual system has been calibrated by this specific human body that makes only our nose visible that way in ordinary circumstances. We can dispense with self-identification. Although it has been largely assumed that bodily IEM is the privilege of bodily experiences, it is shared with some visual experiences.

Visually based bodily IEM sheds new light on the rather intuitive distinction between one's awareness of one's body from the inside and from the outside. It shows that in some circumstances there is no clear-cut epistemological difference between the two as far as bodily IEM is concerned. The fact that some visual experiences are appropriate grounds for bodily IEM also questions the relationship between bodily IEM and the sense of bodily ownership.

3.3 Self-Specificity and Self-Reference

Bodily IEM is the epistemic counterpart of the sense of bodily ownership: there is no judgement of ownership that is sensitive to error through misidentification when it is grounded on experiences of ownership. This is another way of saying that the sense of bodily ownership does not require one to identify the body as one's own. As noted by Martin (1993, p. 210): "This raises the question of how it can be true that we have a sense of our bodies as our bodies, when we don't identify them as such". We shall reply to this question in Part III. But for now let us reverse the relation and ask the following: is the sense of ownership the phenomenological counterpart of bodily IEM? The reply is negative: the lack of self-identification does not suffice for one to experience a sense of bodily ownership. More specifically, the lack of self-identification reveals the self-specificity of the information that is gained, but self-specificity is different from self-reference. Consequently, there are situations in which we do not have a sense of ownership despite the absence of self-identification.

3.3.1 The B-Lander

> The jellyfish lacks all sense of its boundaries with the rest of the world, and has little time for detecting predators [. . .] Although it has sensations which inform it about its body, it is doubtful whether we should think of it as sensing its body as its body. In this case there does not seem to be a useful contrast to be made for the creature itself between its body and the rest of the world.
>
> (Martin, 1993, p. 211)

According to Martin, its body is all there is for the jellyfish: it receives information about its body, and its body only. Yet the jellyfish does not have a sense of bodily ownership. Martin concludes that signals that carry information exclusively about the body that happens to be one's own do not suffice for being aware of the body *qua* one's own.

One way to interpret the jellyfish case is in terms of Perry and Blackburn's (1986) notion of Z-lander. The Z-lander has never moved from Z-land, has never communicated with anyone from another country, and does not even imagine that there exists any other place than Z-land. Perry and Blackburn claim that the Z-lander is so fixated on Z-land that he does not express, or even represent, the spatial location of certain general events. For instance, the Z-lander conceives of rain as a monadic property of time (he thinks that it can rain today or tomorrow, but he never thinks that it can rain here or somewhere else). Therefore, he does not, and even cannot, specify where it is raining. Similarly, one can conceive of a *B-lander* (Body-lander), who is acquainted exclusively with his own body. One may then say that the B-lander is so fixated on his own body that he cannot express or represent whose body hurts or whose body is touched. Since he cannot discriminate his body from what is not his body, he cannot individuate the body that he feels as being his own.

What the thought experiment of the B-lander illustrates is the difference between self-specificity and self-reference, between degree 0 and degree 1 of self-representation (Peacocke, 2014). Self-specific information corresponds to information exclusively about the body that happens to be one's own. It is necessary for being aware of the body *qua* one's own. However, it is not sufficient. In other words, it does not suffice to be aware exclusively of one's body for one to be aware of it *as one's own*. The sense of ownership cannot be reduced to a privileged informational link to one's body. Consequently, as we shall now see, bodily judgements grounded in self-specific information can be immune to error in the absence of the sense of bodily ownership.

3.3.2 Bodily IEM and the Sense of Bodily Ownership

A first example of dissociation between bodily IEM and the sense of bodily ownership can be found in self-specific first-person visual experiences. When one is visually aware of one's body one does not experience it as one's own[7] (Martin, 1995; Brewer, 1995; Bain, 2003):

---

[7] One might be tempted to take the RHI as evidence for the fact that one can visually experience a sense of bodily ownership. However, not only does one see the rubber hand,

when we perceive it from the outside, our body has no indelible stamp of ownership. It appears just as one object among many, although it is one whose features we know very well.   (Brewer, 1995, p. 305)

There is no "stamp of ownership" even in visual experiences that secure bodily IEM. Imagine that your nose has been anaesthetized. Still, you can see it by closing one eye and you cannot doubt that this is your nose. But do you feel ownership towards the nose that you see? Quite the contrary, I would say. The nose from this perspective seems rather unfamiliar. The informational link that secures that this is your own nose does not give rise to the sense of ownership.

Another example of dissociation between bodily IEM and the sense of bodily ownership can be found in the case of Ian Waterman, the deafferented patient. At the beginning of his disease, he could feel warmth on his hand, correctly judge that it was his own hand on the basis of his thermal sensation, and yet feel his hand as alien. More generally, all non-delusional patients, who experience one of their limbs as alien, while still believing that it is their own, reveal that bodily experiences can ground judgements that are immune to error, independently of any phenomenology of ownership.

To conclude, the fact that a perceptual experience carries self-specific information suffices to guarantee bodily IEM, whether the experience is visual or proprioceptive. Our proprioceptive system has been shaped by biological facts, including the invariant fact that the body that we feel is normally our own. We can thus dispense with self-identification when the body is felt. Likewise, our visual system has been shaped by biological facts, including the invariant fact that the body that we see from certain visuo-spatial perspectives is normally our own. We can thus dispense with self-identification when the body is seen in this way. But the lack of self-identification does not equate to the phenomenology of ownership. Therefore, one should not draw consequences about the phenomenology of ownership merely on the basis of considerations about bodily IEM. If one experiences the body part as a part of one's own body, then the judgement of ownership is immune to error through misidentification, but the reverse is not true. If bodily self-ascriptions of the type "my body part is F" are immune to error, it does not follow that one experiences the body part *qua* one's own.

---

one also feels tactile sensations in it. Arguably, the sense of ownership arises from these tactile experiences, and not from the visual ones.

3.4 Conclusion

We have seen that self-specific information about one's body can secure bodily IEM, but that it does not suffice for the sense of bodily ownership. As Martin (1993, p. 212) rightly points out, one needs also to be able to distinguish what is inside from what is outside one's body for one to experience one's body as one's own: "Unlike the jellyfish, we have a sense of ourselves as being bounded and limited objects within a larger space, which can contain other objects". In other words, we are not stuck in B-land. We are aware of its frontiers. We can discriminate B-land from other countries. Does that suffice? Martin thinks so, but we saw in Chapter 2 that even the awareness of bodily boundaries might not suffice. More is needed, but what? In order to answer this question, we shall need to systematically explore the spatiality of bodily experiences. This shall be the object of Part II of this book and it is only in Part III that I shall be able to offer a full account of the sense of bodily ownership.

PART II

Body-Builder

Who am I?
Where am I from?
I'm Antonin Artaud
And since I speak
As I know
In a moment
You'll see my present body
Shatter to pieces
And gather itself
In a thousand notorious
Aspects
A fresh body
In which you'll never
Be able
To forget me.

Artaud (1948)

Part I has left us with a series of puzzles about the sense of bodily ownership, which one can hope to solve only if one has a better understanding of the various aspects of bodily awareness, including the spatiality of bodily experiences, their multimodality, their role in social cognition, and their relation to action. Each chapter can be conceived of as a building block for the general theory of the sense of bodily ownership that I will defend in Part III.

In particular, I have so far taken for granted the notion of bodily experiences (or bodily sensations), simply mentioning the role of information coming from bodily senses. It is now time to offer a better characterization of their nature. Visual experiences remain the main paradigmatic example of perceptual experiences both in philosophy and in cognitive science. Proprioception and touch, on the other hand, have been left largely unexplored.[1] Even when tactile experiences are investigated, it is mainly in the context of haptic exploration, and the focus is exclusively on the perception of the external world with which we are in contact. One of the objectives of Part II is thus to explore the peculiarities of bodily experiences, including their spatial bodily content and their aetiology. In Chapters 4 and 5, I will contrast two main approaches to the spatiality of bodily experiences, a sensorimotor approach and a representationalist approach, and I will argue in favour of the latter. This will allow me to introduce the notion of body map, which will play a central role in my argumentation.

Once I have established the existence of the body map, I will describe its properties. In Chapter 6, I will argue that it does not result only from somatosensory information and that visual information is essential to it. Recent literature in psychology and neuroscience has emphasized the importance of multisensory interactions. Yet the notion of multimodality has become of interest to philosophers only recently (see for instance Bennett and Hill, 2014). I will defend the view that the influence of vision is not an accident, and that it is required by bodily awareness to accomplish its function of reliably tracking bodily states and properties. In Chapter 7, I will consider to what extent the body map is exploited in social cognition and whether it can be said to be interpersonal. Finally, in Chapter 8, I will turn to the relationship between the body map and action. In particular, I will revisit the classic distinction between body schema and body image and show how to best interpret the former in terms of Millikan's notion of pushmi-pullyu representations.

---

[1]   See Massin (2010) and Fulkerson (2014) for their work on touch.

4

Bodily Space

In Part I, I argued that there is something it is like to experience one's body as one's own. I further suggested that the phenomenology of ownership has its source in some fundamental facts about the spatiality of bodily awareness. This then raises the following questions: which facts are relevant and how can they ground the sense of bodily ownership? It is only if we succeed in accounting for the spatiality of bodily experiences that we can hope to solve the puzzles of ownership.

The spatiality of bodily experiences, however, is itself puzzling. Coburn (1966) notes that the pain that I feel in my leg is not felt in the refrigerator when my leg is in the refrigerator. This seems hardly surprising since one cannot feel pain in external objects. But the same reasoning applies even within bodily limits: the pain that I feel in my thumb is not felt in my mouth when my thumb is in my mouth (Block, 1983). This simple example has given rise to a major philosophical debate,[1] but the simple fact that we feel pain to be located in my thumb is already relatively puzzling. How is the spatial content of bodily experiences so rich given the paucity of information carried by the bodily senses? On the one hand, one receives raw somatosensory signals about muscle stretch, tendon tension, joint angle, pressure, temperature, and so forth. On the other hand, one can feel a sensation of touch as being located in the centre of one's right palm on the left side of one's body, for instance. How does one get the latter from the former? A further source of puzzle comes from the fact that bodily sensations can be felt in one's hand,

[1] The lack of spatial transitivity raises questions about the interpretation of the term "in" in the context of the location of bodily sensations. In a nutshell, does it have a spatial meaning or not (Noordhof, 2001; Tye, 2002)? The difficulty with this kind of example, however, is that it combines the peculiarities of the spatiality of bodily experiences in general with the peculiarities of pain itself. For the sake of simplicity, I shall thus focus almost exclusively on tactile sensations.

in one's phantom limb, at the tip of a tool, in a rubber hand, or in a virtual hand, but they cannot be felt nowhere in particular, nor beyond the apparent confines of one's body, or so it seems. Why is this so? Here lie the real mysteries of the spatiality of bodily awareness.

One might conclude with Merleau-Ponty (1945, p. 98) that "ordinary spatial relations do not cross" the experience of the body. But then how should one understand the spatial organization of bodily awareness? All agree that one can report or point to the location at which one feels bodily sensations. Yet as we shall see, there is little agreement on how we should understand such spatial abilities. The objective of this chapter and of Chapter 5 is to explore the grounds of the spatial phenomenology of bodily experiences.

4.1 Are Bodily Sensations Spatial?

Imagine that two pressures of equal intensity are applied on your cheek and on your knee inducing two tactile sensations. In what sense do these two tactile sensations feel different? They guide your attention respectively to two distinct parts of your body, they calibrate your subsequent actions differently, and they ground two distinct beliefs about their respective locations. Yet none of these effects suffices to show that there is a specific spatial phenomenology that is constitutive of your sensations. According to a dominant view in the nineteenth century known as the local sign theory, there is actually no such spatial phenomenology (Lotze, 1888; Wundt, 1897; Titchener, 1908; James, 1890). As James (1890, p. 798) put it, bodily sensations have no "tone as it were which cries to us immediately and without further ado, 'I am here', or 'I am there'". But if there were no spatial information carried by the sensation itself, how could one correctly attribute it to the right part of the body? And could there be bodily sensations that are experienced nowhere in particular?

4.1.1 The Local Sign Theory

Let us go back to our example of feeling touch on the cheek and on the knee. According to the local sign theory, pressure exerted on the knee feels different from pressure applied on the cheek because of the specificity of the area that is touched: the "fleshy" sensation indicates the cheek, whereas the "bony" sensation indicates the knee. These sensations are sometimes described as "auxiliary impressions" (Lotze, 1888, p. 309) or "peculiar qualitative colouring" (Wundt, 1897, p. 105). Although they are non-spatial, they can be taken as *signs* of the locality of the tactile stimulation and they can ground our spatial abilities.

> With sufficient sharpening of the attention we may, confining ourselves to the quality of the feelings alone, entirely abstract from their locality, and yet notice the difference quite as markedly.   (Wundt, 1897, p. 105)

An immediate problem for the local sign theory is that the qualitative differences that it assumes are controversial and many people have denied that they systematically exist (Vesey, 1961; Coburn, 1966; Martin, 1995; Brewer, 1995; O'Shaughnessy, 1980, vol. 1; Holly, 1986; Margolis, 1966). It is true that the density of tactile receptors is not the same all over the surface of the body and there can be phenomenological differences between specific body parts (such as the bony knee and the plump cheek). But there is no non-spatial difference between two sites that are slightly apart on the same limb and as much as I can try, I cannot feel any non-spatial difference between sensations in my left hand and sensations in my right hand, even if "the penalty for error were sudden death", as O'Shaughnessy (1980, vol. 1, p. 157) puts it.

Furthermore, it is one thing to notice a difference—if there is any—between a sensation in the knee and a sensation in the cheek. It is another thing to localize this sensation in the cheek itself. Specific qualitative colouring may be the natural sign of the body site of the sensation: its experience cannot but betray the location of the sensation in the same way that smoke cannot but betray the presence of fire. Nonetheless, the presence of smoke needs to be interpreted on the basis of prior knowledge about the systematic association between smoke and fire. Similarly, the experience of a specific qualitative colouring needs to be interpreted on the basis of prior knowledge about its systematic association with a specific body site. The question then arises: what is at the origin of such spatial knowledge? And in its absence, what do we experience? Free-floating sensations?

4.1.2 Free-Floating Bodily Sensations

Can one feel bodily sensations without at least roughly ascribing them to particular parts of the body? If we accept with the local sign theorists that the local sign needs to be interpreted with the help of further spatial knowledge, then it should be conceivable to feel bodily sensations nowhere in particular when such spatial knowledge is missing. This seems to be true of what Armstrong (1962) calls *bodily feelings*, which include thirst, hunger, and fatigue. Those feelings indeed are not experienced as having specific locations. For example, when one feels thirsty, one does not experience thirst as being located in a particular part of the body (at most, one can feel associated localized sensations of dryness in the throat). By contrast, it seems more difficult, if not impossible, to conceive of tactile sensations that are not localized.

Still, one might suggest that this is the case in early infancy: if spatial knowledge is a prerequisite for being able to localize bodily sensations, then there might be a stage in development in which children have not acquired it and experience free-floating sensations. However, it has been shown that very early on infants are able to localize their sensations (Bremner, 2017). Spatial orienting to (invisible) tactile stimuli is observable in a range of neonatal behaviours. Newborn infants automatically grasp objects in contact with their hands (even when the object is not seen). In addition, infants, as young as 6.5 months old, orient their behaviour to the hand that was touched by looking at it (Bremner et al., 2008). This behaviour should not be interpreted in terms of spinal reflex because it occurs far too late for that, after 1 to 4 seconds after tactile stimulation. It thus seems that some spatial abilities can be found very early on.

One might also expect free-floating sensations in patients with brain lesions if these lesions had impaired the spatial knowledge required for the localization of bodily sensations. No such cases have been described in the literature so far. The closest that one can find is a patient who complained, "I feel you touch me, but I can't tell where it is", but who further qualified his claim: "The touch oozes all through my hand" (Head and Holmes, 1911, p. 139). What this example shows is that it is important to contrast free-floating sensations (aspatial), with those that are *spatially diffuse* (rough-grained localization in regions of the body), and with those that are *mislocalized* (localized in a specific part of the body that nothing touches). Hence, one can be approximate or mistaken in the localization of bodily sensation, but even then the sensation is localized.

Interestingly, when the localization is difficult for one reason or another, one takes more time in detecting touch, suggesting that tactile awareness requires spatial processing to be completed (Gallace and

Spence, 2010). For instance, when one wears optical prisms that distort the direction of light rays by a constant angle, one sees one's hand at a location distinct from the location at which one feels it to be. If the hand is touched, while one is looking at it with prisms, one takes longer to detect tactile stimulation (Folegatti et al., 2009). Roughly speaking, as long as no solution to the conflict has been found, one is not aware of the touch. A similar slowing-down of tactile detection has been found in the RHI, in which there is an apparent conflict between the visual information about the location of the rubber hand and the proprioceptive information about the location of one's hand (Moseley et al., 2008). Therefore, there seems to be a systematic relationship between tactile sensations and localization: the less spatially determinate, the more time it takes to become aware of touch.

None of these findings suffices to show the impossibility of free-floating sensations. Nonetheless, they suggest that to be aware of touch is to be aware of touch as being located, though not necessarily correctly. The view that I defend is thus that it is part of the identity conditions of bodily sensations that a specific sensation feels to be located in a specific part of the body.

4.1.3 The Duality of Spatiality

Not only are bodily sensations essentially spatial but they also have a dual spatial content. As O'Shaughnessy (1980, vol. 1) claims, bodily sensations are "sensations-at-a-part-of-body-at-a-point-in-body-relative-space":

Look up at the ceiling while your right arm is moved to the right. Attempt to recapture the original uninterpreted experience. Try to undo the lessons of time! Now somehow one cannot shake off the belief that the arm has gone to the right. Well, try again. Hard. Attempt to transfer all attention away from the limb and onto the feelings. Scrutinise them. Now tell us about them. Then you cannot in doing so help mentioning that they were in your arm, which was over to the right, poised like such and such, near to such and such a part of the body.

(O'Shaughnessy, 1980, vol. 1, p. 157)

One way to interpret O'Shaughnessy's claim is that bodily sensations have two types of felt location, which I call *bodily location* and *egocentric location*.[2] In O'Shaughnessy's example, the bodily location is my right

---

[2] Also called A-location and B-location (Bermúdez, 2005).

arm, whereas its egocentric location is over to the right. The bodily location, which is given in a skin-based frame of reference, is the location in a specific part of the body, no matter where the body part is located. When I move my hand, the bodily location does not change, unlike the egocentric location. The egocentric location indeed depends on bodily posture, and it is thus given within an egocentric frame of reference: it is the location relative to the other parts of the body (eye-centred, head-centred, trunk-centred, and limb-centred) at the time of the sensation.

The hypothesis of dual spatial content is true not only of proprioceptive experiences but also of tactile experiences, although the awareness of egocentric location is less phenomenologically salient (for review, see Badde and Heed, 2016). Consider the Japanese illusion. Cross your wrists, your hands clasped with thumbs down. Then turn your hands in towards you until your fingers point upward. If now I touch one of your fingers, you will have difficulty not only in moving the finger that was touched but also in reporting which finger it was. If bodily sensations had only bodily location, then this complex posture should make no difference. A touch on the right index finger remains on the right index finger no matter where the finger is located. The difficulty that you experience in the Japanese illusion, however, shows that (i) the relative location of body parts matters and (ii) bodily location can be perceived as being in conflict with egocentric location (for instance, a right hand on the left side of the body).[3]

Likewise pain has both bodily location and egocentric location, and the latter can take priority over the former, as illustrated by the following study using the thermal grill illusion. In this illusion, a sensation of painful heat is elicited by touching interlaced harmless levels of warm and cool bars to the skin (for instance, index finger warm, middle finger cold, ring finger warm). What happens when the fingers are crossed? If the illusion had only relied on the bodily location of the thermal stimulations, then it should change nothing. However, it was found that the illusion exploited egocentric location, that is, the relative location of the stimulated fingers. With the middle finger crossed over the index finger for instance, the cold stimulation must now be applied to the index

---

[3] The effect of crossing one's hands on temporal order judgement described in Chapter 2 is also driven by the interference between egocentric and bodily locations (Yamamoto and Kitazawa, 2001a).

finger (which is in between the middle finger and the ring finger) for the illusion of painful heat to work (Marotta et al., 2015).

Further evidence of the duality of the spatial content can be found in deafferented patients. As described in Chapter 2, these patients do not receive proprioceptive and tactile information about their body below the neck. Yet they can feel pain and thermal sensations. Hence, if you pour cold water on their right hand while they are blindfolded, they are able to localize their sensation in their right hand, although they have no idea where their right hand is. In other words, the thermal sensation has a bodily location without an egocentric location. As Brewer concludes (1995, p. 301): "So G.L. [one of the deafferented patients] has knowledge of the location of those very few bodily sensations of which she is aware that is relatively different in kind and content from our own".

Hence, not only do bodily sensations have a spatial content but they have a dual one. I shall now consider whether one can successfully account for this dual spatial content in sensorimotor terms. From Merleau-Ponty (1945) and Anscombe (1962) to the more recent enactivist theorists, it has indeed been assumed that the locations of bodily sensations are constitutively linked with dispositions to direct actions at these locations.

4.2 A Sensorimotor Approach

According to Merleau-Ponty, bodily space is a space of actions. They need not be performed but can remain virtual. In his words, the lived body consists in an "I can" (1945, p. 137). For instance, he claims that phantom limbs are the consequence of the preserved readiness to move the amputated limbs.[4] More generally, Merleau-Ponty argues in favour of the constitutive role of action for consciousness. As such, he may be considered as one of the precursors of the recent sensorimotor approach, according to which conscious experiences are inseparable from bodily activities (Legrand, 2006; Siewert, 2005; Noë, 2004; Hurley, 1998; Thompson, 2005; O'Regan and Noë, 2001; O'Regan, 2011). On this view, what we feel is determined by what we do and what we know how to do. Proponents of the sensorimotor approach thus argue for a

---

[4] However, many patients feel that they cannot move their phantom limbs.

relation of constitutive interdependence between perception and action. Perception is not merely a means to action, nor is action only a means to perception. Instead, perceptual experiences, and especially spatial experiences, constitutively depend on our ability to keep track of the interdependence between sensory inputs and motor outputs, this ability being grounded in the procedural knowledge of how the way in which we move affects the sensory signals that we receive (or how the movement of objects affects these signals). Perceptual experiences are said to be inseparable from the perceiver's bodily activities. Although it is vision that has received the most attention from enactivism, the sensorimotor approach is the most plausible when applied to tactile experiences. Noë (2004, p. 1) even begins his book *Action in Perception* by claiming "All perception is touch-like":

> Touch acquires spatial content—comes to represent spatial qualities—thanks to the ways touch is linked to movement and to our implicit understanding of the relevant tactile-motor dependencies governing our interaction with objects.
>
> (Noë, 2004, p. 205)

Noë has in mind haptic touch, which consists in exploratory activities, but could the sensorimotor approach account for passive touch, and even more so for bodily experiences in general? Here I shall not review the many objections that have been offered against the sensorimotor approach (see Block, 2005, for instance), but focus on those that apply to the sensorimotor account of bodily spatiality.

4.2.1 Tactile-Motor Laws

> When something tickles you on your arm, you can move your other hand and scratch the tickled location [. . .] If you actively move the part of the body being touched, the incoming sensory stimulation will change [. . .] What we mean by feeling something in a particular place in our bodies is precisely that certain sensorimotor laws apply.
>
> (O'Regan, 2011, pp. 157–8)

As basic and uncontroversial as the tactile-motor laws suggested by O'Regan are, they are problematic if they are supposed to account for tactile awareness. For instance, the movement of separating the body part from the object by moving it stops any tactile experience, but it does so whether I feel a rough texture or a smooth one. Hence, it cannot be an individuative condition of *what* I feel. But can it individuate at least *where* I feel it, which is at the core of our interest? Imagine that a spider lands on the back of my right hand. I feel its light contact. According to the sensorimotor view, feeling its contact on my right hand consists in knowing that if I move my left hand towards the location of my right hand and remove the spider, I will stop the tactile stimulation. The problem is that this law may account for egocentric location (the relative location of the two hands), but not for bodily location. Now let us imagine that I wave my right hand in an unsuccessful attempt to dislodge the spider. My right hand is no longer at the same location and thus the path that my left hand must take to reach it has changed. One may then say that the tactile-motor law has changed. And indeed the egocentric location of the sensation has changed, but not its bodily location: I still feel the contact on the back on my right hand. When I move my hand, I experience my hand moving, I do not experience the sensation moving. Hence, the tactile-motor law, which depends on the egocentric location of the sensation, cannot account for its bodily location.

A proponent of the sensorimotor approach might then deny that sensorimotor laws take into account the precise pathway between the two hands. They encode only the general fact that reaching for the hand that is touched with the other hand will stop the tactile signal. But if sensorimotor laws are at this level of generality, then they fail to account for egocentric spatial content, without necessarily gaining a clear account of bodily location. Tactile-motor laws are body-part specific. It is only if I reach for my right hand, and not for any other limb, that I can expect the tactile signal to stop. But how do I know that it is my right hand that must be the target of my movement? There is a threat of circularity, which is similar to the one highlighted by Jacob (2008) in the case of vision. The sensorimotor account claims that visuo-motor knowledge consists in the knowledge of how one's movements towards a visually seen object affect the way one perceives the object. However, Jacob notices that in order to act upon an object, one needs first to be able to single it out and this cannot be done on the basis of visuo-motor expectancy for fear of circularity. Likewise, tactile-motor knowledge requires a prior and inde-pendent way of singling out the relevant body part that is in contact with the object in order to avoid a circular account of bodily spatial content.

Other sensorimotor theories do not fare any better. For instance, Noë (2004) claims that the content of perceptual experience consists in the procedural knowledge that provides access to objective properties based on the relation between the experience of perspectival properties and bodily activities. He gives the example of the table that is round (object-ive property of the stimulus independent of the perceiver), but which appears elliptical to you (perspectival property, the way the stimulus appears to the perceiver). According to Noë, your visual experience represents the table as round because you know how its perspectival properties will change depending on how you move relative to the table (for example, if you get closer to it). However, it seems difficult to apply this account to the spatial content of bodily experiences. When I feel the spider on my hand, it does not seem to make sense to distinguish between a perspectival bodily property and an objective bodily property, if perspectival properties depend on the spatial relation between the perceiver and the perceived object. This is so because one cannot change the spatial conditions under which one's own body is felt from the inside:

> Bodily sensations, even if they could be moved about, would be moved about within the phenomenal body [...] there are no varying perspectives to be had on them, no varying approaches that we can make to them by actively moving our sense organs in relation to them. (Smith, 2002, p. 153)

And if one cannot change the spatial conditions, then there is no corres-ponding tactile-motor knowledge of the way the perspectival properties change depending on the way one moves relative to one's body.

Even if the tactile-motor laws outlined above were satisfactory, they would not apply to *instantaneous* touch. Imagine now that the spider has landed only very briefly on my hand before jumping elsewhere. Then I cannot remove the spider because it is already gone. What I do will make no tactile difference because the signal has already stopped. More generally, there cannot be any sensorimotor contingencies when percep-tion is so temporary that it does not afford the time to act, and thus to have sensory consequences of one's actions. Most sensorimotor laws require enduring perceptual experiences. One may then be tempted by a purely motor (instead of a *sensori*motor) account of bodily experi-ences. For instance, you can still know how to go to the place at which there used to be a restaurant that you liked, although it is no longer there. Similarly, I can still act towards the location of the tactile stimulus, although the stimulus is no longer available. One may then suggest that the spatial content of tactile experiences is determined by the procedural knowledge of how to get to the bodily location of the tactile stimulation, knowledge close to what Siewert (2005) calls *bodily know-how* (i.e. knowing how to reach or move the body part that instantiates the bodily property). Accounts in terms of bodily know-how sound like a plausible working hypothesis for the sensorimotor approach. However, this type of sensorimotor account is challenged by a series of empirical results that reveal that contrary to what the sensorimotor approach predicts, tactile experiences are separable from the perceiver's motor knowledge.[5]

4.2.2 Dissociation between Bodily Know-How and Bodily Experiences

> But, I don't understand that. You put something there; I do not feel anything and yet I got there with my finger. How does that happen?
>
> (Paillard et al., 1983, p. 550)

This patient, who was blindfolded, was amazed at her own ability to point to the site at which she was touched on her hand while she felt absolutely no sensation of having been touched. She suffered from what is called "numbsense" in the neuropsychological literature (also called "blind touch"). Following cortical or subcortical lesions, patients with numbsense become completely anaesthetized on their right side. They are not able to detect, localize, and describe tactile stimuli on their right arm. Nor are they able to indicate on a picture of an arm where the stimulus is applied, even in a forced choice condition (Rossetti et al., 1995). Yet despite their apparent numbness, they are able to guide their opposite hand towards the approximate site at which they are touched when so instructed, and to their own surprise.[6] Likewise, although they

---

[5] Likewise, one of the most recurrent and powerful empirical counterarguments provided against the sensorimotor approach to vision comes from the neuropsychological dissociation between optic ataxia (preserved visual judgements with impaired visually guided actions) and visual agnosia (preserved visually guided actions with impaired visual judgements), and from the literature on visual illusions (action can be immune to visual illusions, such as the Müller-Lyer illusion, the Ponzo illusion, the Titchener illusion, and the hollow face illusion). These results have been taken as evidence that the content of visual experiences may be at variance with the content of action-oriented vision (for a review, see Jacob and Jeannerod, 2003).

[6] It is worth mentioning that they are not perfectly accurate in their pointing responses (Medina and Coslett, 2016). However, if the sensorimotor view were correct, one would expect patients to have inaccurate bodily experiences instead of a complete lack of sensation.

are unaware of their arm location, they can accurately reach the position of their arm. Hence, they are able to motorically localize their hand, indicating thus that they master some bodily know-how. Yet they cannot verbally localize their hand or their tactile sensations, since they feel nothing. The case of numbsense shows that bodily know-how (and its exercise) is not a sufficient condition of tactile and proprioceptive experiences. Even more surprisingly, their preserved bodily know-how does not even help patients with numbsense to localize their sensations: they are no better in judging the location of their sensation when they are acting at the same time; instead they become equally bad in the verbal and in the motor tasks (Rossetti et al., 1995).

One may reply that bodily know-how can ground spatial ascription of bodily experiences, but only if it is combined with the presence of some raw sensations (Noë, 2005). In its absence, as in numbsense, it can play no role. However, this answer is undermined by the case of allochiria. Patients with allochiria feel bodily sensations but systematically mislocalize them to the contralateral side. For example, the patient RB had perfect understanding of left and right, she did not suffer from spatial neglect, but when asked where the experimenter touched her, she was 100 per cent inaccurate, always misattributing the touch to the other side of her body. Furthermore, she spontaneously reported feeling pain in her right knee but she pointed and rubbed her left knee (Venneri et al., 2012). The case of RB illustrates how bodily know-how can be at odds with the spatial content of bodily experiences, even in the presence of a raw conscious sensation.

Further dissociation between bodily experiences and bodily know-how can be found in the following study (Anema et al., 2009). KE and JO can both consciously feel a touch. In addition, they can accurately report the location of a visual target and point to it. Yet when asked to point to where they were touched, they displayed a surprising double dissociation. It was found that JO failed to point accurately on a pictorial representation of her hand (i.e. abstract and detached perceptual report), whereas she was able to point accurately to her own hand (i.e. tactile-motor task). Hence, her intact bodily know-how did not suffice for providing her tactile experience with accurate spatial content. By contrast, KE pointed correctly to the hand drawing, but failed to point accurately to his own hand. Although KE knew where he was touched, he was unable to get to the right location of the touch on his hand. In his case, there was no bodily know-how that could provide the spatial content of his tactile experience.[7] The tactile spatial content must have had a different ground. This indicates that bodily know-how is not a necessary condition for spatial ascription.

The dissociation between the two types of spatial information specific to each system may be made even more salient in the context of bodily hallucinations (Pitron and Vignemont, 2017). Consider, for instance, the Alice in Wonderland syndrome. Its name finds its origin in Lewis Carroll's (1865) novel, in which Alice perceives her body growing bigger or shrinking after she ate and drank magic food. Similarly, during Alice in Wonderland hallucinations, the perception of the size of one body part is altered: patients experience their limbs as smaller or larger than they really are. The sensorimotor view might explain their feelings by erroneous bodily know-how. One should then expect the Alice in Wonderland syndrome to be associated with a range of motor deficits. However, to our knowledge no motor disorder has been reported in the clinical descriptions of the syndrome and it is classically described as a purely perceptual distortion of the representation of one's bodily metrics (Lippman, 1952; Todd, 1955). For instance, a patient reported that she suddenly experienced her legs as being very short while she was walking but the fact is that she walked normally (Lippman, 1952). Hence, there was a complete disconnection between her bodily awareness and her bodily know-how: her hallucination had no impact on her movements and her normal steps did not cancel the illusion of having short legs.

Unfortunately motor performance has never been explicitly investigated in bodily hallucinations. Let us thus turn to bodily illusions in healthy individuals. Again the sensorimotor view should account for these illusions in terms of erroneous bodily know-how. The explanation might run as follows for the RHI: participants localize their hand closer to the location of the rubber hand because they are misleadingly induced

---

[7] The patient KE had no motor deficit, as shown by his preserved ability to point to a hand map or to neutral visual targets. His results could be explained neither by a proprioceptive deficit nor by a memory deficit. In the study, the patients' hands were moved passively to a different location and the patients were asked after two seconds to return their arms to the previously held position. KE's performance in this proprioceptive task was similar to JO's performance. It thus seems plausible to assume a deficit of bodily know-how in KE, which explains his poor motor performance when pointing to his own hand.

to expect that if they reach the location close to the rubber hand, they will touch their own hand. However, this explanation cannot account for the following facts (Kammers et al., 2009a; see also Kammers et al., 2006 for a dissociation in kinaesthetic illusions).[8] Participants accurately directed their opposite hand to the real location of their own hand that was touched, and not to the illusory location that they reported. Their reaching movements were not sensitive to the spatial illusion. Similarly, when they performed the reverse movement (i.e. directing their touched hand towards their opposite hand), they had the right know-how of where their hand was. The complete absence of illusion was also confirmed in a bimanual task, where participants had to grasp a stick in front of them. Again, their bodily movements revealed accurate spatial information about the relationship between their two hands as well as the use of bodily know-how of their correct locations. When participants were asked a second time to make a perceptual judgement about the location of their touched hand after having moved, they were still sensitive to the RHI, and they still localized their hand as being closer to the location of the rubber hand than it was (although less so). Hence, the RHI cannot be explained by inaccurate bodily know-how. The illusory spatial content of the bodily experience is not determined by the illusion-immune bodily know-how, which reveals an accurate sense of the position of the touched hand.

To sum up, in both patients and healthy individuals, the spatial content of tactile and proprioceptive experiences can be dissociated from the spatial information encoded in bodily know-how used to guide reaching and pointing movements so that one can be inaccurate in one and not the other, and vice versa (see Table 4.1). Numbsense patients even illustrate the fact that it is possible to have preserved know-how with no associated tactile sensation. Hence, it seems that bodily know-how, as recruited by reaching, pointing, and grasping movements, is neither necessary nor sufficient for bodily experiences (Vignemont, 2011).

How do the sensorimotor advocates respond to these cases? O'Regan (2011) argues that the notion of bodily know-how is too narrow. Instead, he suggests that the spatial content of bodily experiences consists in a list

---

[8] We shall examine in greater detail the effect of the RHI on actions in Chapters 8 and 9.

Table 4.1. Dissociations between bodily experiences and bodily know-how

| | Bodily experiences | | Bodily know-how |
	Conscious detection	Felt location	
Numbsense	Impaired	Incorrect	Correct
Allochiria	Preserved	Incorrect	Correct
Patient JO	Preserved	Incorrect	Correct
Patient KE	Preserved	Correct	Incorrect
RHI	Preserved	Incorrect	Correct

of multiple sensorimotor laws in various sensory modalities (if I turn my head towards the location that is touched, I will see what is touching me, for instance). If the list of sensorimotor laws is too inclusive, however, the risk is that the theory could never be falsified. Noë (2010), on the other hand, claims to be interested in what perceptual experiences are grounded in, and not in what they are for. Reaching or grasping movements, on his view, are mere "practical consequences" of bodily experiences. Indeed, not all relations between touch and action are of interest to the sensorimotor view, but only those that are said to be non-instrumental (Hurley, 1998). But why should perceptual experiences not be grounded in what they are for? And what are the criteria that can be employed to distinguish between the "right" and the "wrong" types of procedural knowledge?

4.3 Conclusion

Although they seem to be intimately linked, one cannot merely assert that action is crucial to the spatiality of bodily experiences. One needs to make more specific claims. According to the most radical interpretation, bodily experiences consist in sensorimotor laws. Sensorimotor theories, however, cannot easily explain dissociations between action and bodily experiences. Nor can they provide satisfying sensorimotor laws that account for the spatial ascription of bodily experiences. I shall thus turn to a different approach to bodily awareness, a representationalist one.

Whereas the sensorimotor approach focuses on egocentric location and highlights the importance of action for the spatiality of bodily experiences, the representationalist approach focuses on bodily location and appeals to mental representations of the body (Vignemont and Massin, 2015).

5

The Body Map Theory

Imagine that you have a cross on a map indicating, "You are here", but the map is blank: there is no reference point and no orientation. The cross is then of little interest. It is only when you put tracing paper with information drawn on it over the blank map that you can know where you are. Likewise, if the only spatial information that one had came from somatosensory signals, then bodily sensations would be experienced as located at isolated body points, as mere "here" or "there". This is more than what the local sign theory assumes but this is still relatively poor. To achieve a rich spatial content, I will argue that one needs a map of the body that is a representation of the enduring properties of the body, including its configuration and its dimensions. This representation of the body can be decoupled from the biological body leading the subject to experience sensations not only in phantom limbs but also in tools that bear little visual resemblance with the body. Does this entail that there is almost no limit to the malleability of this body representation? Or does this entail that bodily sensations can be felt even beyond the boundaries fixed by this body representation?

5.1 A Representationalist Approach

In his seminal paper "Sight and Touch", Martin (1992) highlights structural differences between visual and tactile experiences. He compares the following two examples: seeing a ring-shaped object, as a Polo mint, and grasping a glass in one's hand. In both cases, one is aware of the circular shape but the way the shape is presented differs. I see the Polo mint on the background given by my visual field (i.e. spatial array of visual impressions). Within my visual field, I am aware of the ring of the mint, but also of the hole inside. By contrast, when I feel the glass between my fingers, there is no directly comparable tactual field, or at least it does not play the same role as the visual field. Indeed what spatially organizes my tactile sensations is beyond touch itself: it involves being aware of the glass within a space that is not tactually perceived. Instead, the tactile awareness of the circular shape is grounded in the awareness of the body that holds it. The question is: what is the basis of this "bodily field"?

One needs proprioceptive information about the location of the fingers that hold the glass but in order to interpret it, one needs further spatial information. For example, the fact that we have five fingers, that they are cylinder-shaped, that they are of a certain length, and that they are next to each other in a certain order cannot be easily derived from signals about muscle stretch, tendon tension, and joint angle. More generally, the bodily senses only directly carry a limited scope of spatial information, and in particular they do not directly inform us about the shape of the various parts of the body, their spatial configuration, and their size. In order to compensate for the insufficiencies of the bodily senses, I thus propose that bodily information needs to be structured by a representation of the configuration and metrics of the body segments. I shall call this view the *body map theory*.

5.1.1 The Body Map

Some early versions of the body map theory can be traced back to Bonnier (1905), who first used the term of body schema to refer to the spatial organization of bodily sensations. In 1911, the neurologists Head and Holmes also posited the existence of a schematic model of the skin surface of the body used for localizing bodily sensations, which is impaired in some brain-lesioned patients: "This faculty of localization is evidently associated with the existence of another schema or model of the surface of our bodies" (Head and Holmes, 1911, p. 187). It has given rise since then to a long tradition in psychology and neuropsychology (Medina and Coslett, 2016; Schilder, 1935; Longo et al., 2009b; Longo and Haggard, 2010; Berlucchi and Aglioti, 2010; and Brecht, 2017 among others). This view is also well developed in philosophy by O'Shaughnessy (1980, 1995), according to which a long-term body image structures all our bodily experiences. On his view, the spatial content of bodily experiences has the following dual specification:

At instant t1 one seems to be aware of a flexed arm because in general (and in fact over a period of decades) one takes oneself to be a being endowed with an arm which can adopt postures like stretched, flexed, etc.; and because of the operation of postural sensations, etc., at t1.   (O'Shaughnessy, 1995, p. 184)

For the sake of simplicity, I will use the term of *body map*, which is less theoretically laden than the notions of body schema and body image (for discussion of these notions see Chapter 8). But what is this body map? Here one may propose a hierarchical model of body representations on the model of Marr's three-step theory of visual perception: the primal sketch, the 2.5D sketch, and the 3D sketch (Marr, 1982). At each step, more information is extracted from the original visual input until it switches from a viewer-centred perspective to an object-centred perspective. Recent work on body representations suggests similar serial processing and that at each stage, the representation of the body gains in complexity and spatial richness (Longo, 2017; Haggard et al., 2017). By body map, I do not refer to the primal body sketch at the level of the primary somatosensory area (SI), also known as the "Homunculus" (Penfield and Rasmussen, 1950; Blankenburg et al., 2003). The Homunculus is indeed a distorted representation of the body, and thus it cannot play the role of supplying information about bodily metrics and shape (Medina and Coslett, 2016). Although it generally follows natural anatomical divisions, associating cortical areas to specific parts of the body, some are over-represented (e.g. hand), whereas others are under-represented (e.g. torso). Furthermore, the Homunculus does not represent anatomical contiguity (e.g. the hand-specific area is next to the face-specific area). Instead, by body map, I refer to a kind of 3D sketch, which is at a higher level of integration (integration among body parts and among multiple sensory sources), and whose function is to carry information reliably about the spatial organization and dimensions of the body and its segments, in order to organize spatially the information coming from the bodily senses.

It is thanks to the body map that sensations are experienced as being at more than an isolated body point: if you are touched on the right index finger and on the right middle finger, you can feel these sensations to be both located *on your right hand*. It is also thanks to the body map that one can experience sensations in the complete absence of somatosensory input. In the Introduction, we saw that we sometimes feel our mouth inflated when we are anaesthetized by the dentist. It has been shown indeed that when the sensory inputs from the lips and front teeth are fully blocked, subjects can report feeling their lips and teeth increasing in size by as much as 100 per cent (Gandevia and Phegan, 1999; Türker et al., 2005). This illusion cannot be explained by information coming from bodily senses since they are completely blocked; it can be explained only by the presence of an off-line representation of the body, namely the body map. Finally, it is thanks to the body map that one can feel sensations in phantom limbs or in tools. Since the body map represents the body, it can misrepresent it. When it happens, as is the case after amputation or after tool use, the bodily location of the sensation is then altered. Consequently, you can feel pain to be located in your phantom hand as long as the body map still includes the missing hand or you can feel resistance at the tip of the cane if the body map has incorporated the tool. Proof of such incorporation is to be found in the alteration of the representation of the size of the limb. For instance, tool use temporarily modifies the perceived size of the agent's arm (Cardinali et al., 2009a; Sposito et al. 2012), whereas amputees overestimate the length of their residual limb as the result of wearing their prosthesis (McDonnell et al., 1989; Canzoneri et al., 2013a).

The body map is thus the background on which bodily sensations are experienced, their spatial frame of reference. To some extent, it plays the role that the visual field plays for visual experiences but there is an important difference: the visual field is always perspectival in the sense that it displays what is seen from the spatial perspective of the subject, whereas the body map is not perspectival. The body is not represented from a specific spatial viewpoint, as already noted in Chapter 4. I suggest understanding the notion of body map as a kind of scenario in Peacocke's (1992) sense. A scenario is a nonconceptual, analogue, unit-free, spatial representation that uses a set of labelled axes in relation to body parts. Peacocke describes scenarios that depicts the subject's environment, but arguably, one can have a scenario that depicts the subject's body, with similar properties. In a nutshell, the body map is a scenario whose nonconceptual content is organized around body parts (e.g. hand, foot) and by spatial relations, such as back/front, left/right, and up/down. More shall be said about its properties in the following chapters.

5.1.2 Bodily Location and Egocentric Location

In Chapter 4 we saw that the sensorimotor approach is primarily concerned with egocentric location. By contrast, the body map theory is

ulder, experiencing sensations
in her shoulder. When she
ry hand, she felt relief from the
orted seeing her supernumerary
hite", and "transparent" (Khateb
ity in somatosensory, motor, and
the supernumerary hand confirmed

be confirmed by a version of the RHI,
hand illusion. Whereas in the classic
hand remains hidden from sight, in the
one can see both one's hand and the
Guterstam et al., 2011). What is interesting
feeling sensations in their hand and in the
, when asked whether it seems as if they had
ee. In another version of the RHI, participants
hands placed side by side above their biological
en behind the table (Ehrsson, 2009; see also
olegatti et al., 2012). After synchronous stroking
the two rubber hands and the participants' hand),
touch on both rubber hands.

tive Model

studies support the liberal model of the body map?
ere are alternative explanations of supernumerary limbs
atible with a more conservative model. A first question is
he hands (phantom hands, rubber hands, and biological
epresented together within the body map. It has been argued
classic RHI, the embodiment of the rubber hand is at the cost
odiment of the biological hand (for discussion, see Lane et al.
he fact is that participants deny that they experience having
nds (Longo et al., 2008). This seems to be confirmed by the
that there is a drop in temperature of the biological hand only
synchronous condition (Moseley et al., 2008) but this finding ha
difficult to replicate and the drop in temperature may be found i
synchronous and asynchronous conditions (Rohde et al., 2013
re interestingly, it has been shown a decrease in motor excitabilit
the biological hand only in the synchronous condition, which may b

primarily concerned with bodily location. The body map indeed represents the structural organization of the various segments of the body independently of current posture. Nonetheless, the body map also contributes to fix egocentric locations. One needs to know the configuration and the metrics of the body in order to know the relative position of all its segments. For instance, when your hand is stretched, the distance between the tip of your fingers depends on their size. Similarly, your arms can be crossed or not with the very same joint angles depending on their size and on the width of your shoulders. Egocentric locations are ultimately grounded in the body map. Consequently, when the body map is altered, not only is the bodily location altered but also the egocentric location. This is well illustrated by the study with the crossed sticks described in Chapter 2: participants have difficulty in judging which stick is stimulated first when the sticks (but not their hands) are crossed (Yamamoto and Kitazawa, 2001b). The difficulty that the participants experience results from the fact that the egocentric location is computed relative to the sticks, which are crossed, rather than relative to the hands, which are uncrossed. In other words, the sticks are represented within the body map and their embodiment has displaced the egocentric location of the sensation. Another example of disruption of egocentric location can be found in patients with autotopagnosia, whose body map is disturbed. In one study, a patient was asked to construct a pictorial image of the body and of the face with pre-cut puzzle pieces (Guariglia et al., 2002). She produced an odd image of the face with the nose next to two ears on each side next to the eyes. Yet she was happy with the result, demonstrating she was unable to judge the oddity of her work. Interestingly, because of the distortion of their body map, patients with autotopagnosia are unable to correctly point to where they are touched[1]:

When she was requested to show where her elbow was, she looked at her body, searching for the elbow and repeating "yeah elbow, the elbow must be over here", while she was pointing near to the wrist and along the arm.

(Sirigu et al., 1991, p. 631)

---

[1] It is worth noting that autotopagnosic patients do not experience free-floating sensations, they only mislocalize them.

Another consequence of the dependence of egocentric location on the body map is that one cannot experience egocentric location without having information about bodily location (Brewer, 1995). Imagine that I touch your left hand, which is on the left side of your body. If you had experienced only the egocentric location of the tactile sensation, then you should experience the sensation somewhere on the left. Now you cross your hands. You should then experience the sensation somewhere on the right. You should thus experience the sensation moving with no fixed spatial location. As seen in our discussion of the sensorimotor views, this sounds hardly plausible. The reverse dissociation, however, is found in deafferented patients, who can experience bodily location without egocentric location. Such patients localize bodily sensations in their right hand although they can have no idea where their right hand is in the egocentric frame.

To recapitulate, bodily sensations are experienced as being located in a particular part of the body in a particular region of the world thanks to the spatial information conveyed by the bodily senses, but this information does not suffice to fully account for the spatiality of bodily experiences. More is needed for them to acquire their relatively rich spatial content, and more particularly the body map is needed to provide the set of loci with respect to which the location of bodily sensations can be defined.

5.2 The Rubber Band Hypothesis

According to the body map hypothesis, the spatial content of bodily experiences is shaped by the body map, which can be distorted, and includes extraneous objects (for a review, see Holmes and Spence, 2005). But are there limits to the malleability of the body map, and thus to where one can feel bodily sensations? The body map may sometimes seem so plastic that one may wonder what biological constraints, if any, it must conform to. Consider the following example from O'Shaughnessy:

The hypothesis is that if in general one took oneself to be (say) octopus-shaped instead, then despite having a human shape and despite the presence of posture-caused phenomena like sensations of posture, one could not have the experience of seeming to be in the presence of a flexed (very roughly) arm-shaped thing.

(O'Shaughnessy, 1995, p. 184)

taken as evidence of at least partial disembodiment (Della Gatta et al., 2016). It might thus be that one cannot so easily experience three hands. Even in supernumerary limbs, one may wonder about the timing of the subject's experience: do patients *simultaneously* experience the super-numerary phantom limbs and the biological limbs? In the RHI with two rubber hands, their incorporation is never assessed simultaneously. It is then possible that participants rapidly switch their attention from one to the other, and that the two rubber hands are never simultaneously represented in the body map. In addition, one may speculate that in the case of supernumerary phantom limbs, the biological limbs, which are almost systematically paralysed with severe somatosensory loss, are simply not included in the body map, but only visually represented. Alternatively, if there are several types of body map, as I shall argue in more detail in Chapter 8, it may be that the two biological limbs are represented in one type of body map, and the supernumerary phantom limbs, which sometimes feel as alien or anarchic, in another type of body map (Weinstein et al., 1954; McGonigle et al., 2002). In all these various interpretations there is no level at which three or four hands are represented all together.[2]

None of the explanations suggested above, however, can account for the supernumerary limb illusion (or for any situations in which one simultaneously feels sensations in extra limbs). Since participants report sensations in the rubber hand and in the biological hand, it may then seem that both must be part of the body map. Furthermore, participants were explicitly asked if they felt sensations in both the rubber and the biological hands *at the same time*, and they agreed. But does this suffice to show that the body map represents two left hands? Not necessarily. The body map is not an online perceptual representation of the body; instead, it may be compared to a stored template of bodily configuration, which is used as a frame of reference to spatially organize bodily experi-ences. When one localizes the touch, one localizes it on the left hand, as it is represented in the body map. In the illusion, it just happens that

---

[2] Interestingly, patients very rarely experience more than a pair of extra hands or legs (Curt et al., 2011; Khateb et al., 2009; Staub et al., 2006; Vuilleumier et al., 1997). When they do so (for instance, up to six heads and five bodies, cf. Weinstein et al., 1954), they are also suffering from delusion of reduplication, which seems to reflect a more general disorder that is not specifically bodily.

two distinct hands, the biological hand and the rubber one, meet the description of the left hand as represented in the body map. Roughly speaking, they concur to be *the* left hand (Guterstam et al., 2011). And if no cue allows one to decide which is the best candidate for being the left hand, then they are both conceived of as being the left hand. The frame of reference given by the body map, which represents a single left hand, is used twice to structure two distinct bodily experiences.

To conclude, the body map may be conceived on the model of a *rubber band* (Vignemont and Farnè, 2010). Rubber bands have three interesting properties: (i) you can stretch them, but only to a certain degree; (ii) they cannot shrink beyond their size by default; (iii) once you relieve the tension, they regain their default size. These three properties are to some extent exhibited by the body map. Firstly, one can literally stretch the body map, as in the Pinocchio illusion (in which one can feel one's nose to be growing by as much as 30 cm), or after tool use. But it does not mean that there is no limit to the extension of the body map. In particular, we have just seen that it probably does not represent more than two hands. Secondly, the fact that it can still represent amputated limbs may indicate that it does not "shrink" easily. This is confirmed by some evidence showing that it adjusts more readily to increase in limb size than to decrease (Vignemont et al., 2005a; Pavani and Zampini, 2007; Bernardi et al., 2013), a fact that can be easily explained by onto-genetic considerations. Finally, the third property of the rubber band seems to apply especially to the characterization of the body map. A large part of the experimental investigation of body representations has focused on the ways in which tool use stretches the body map but one should not forget that each time we put down a tool we need to revert to our original body map. We thus automatically recalibrate the correct size of our limbs a couple of minutes after tool use. How do we achieve this recalibration? Given the number of times we drop off tools in everyday life, it does not seem parsimonious to assume that the size of one's arms is recomputed every time one relinquishes a tool. It thus seems likely that there is a default body map, equivalent to the default size of the rubber band. I will come back to this notion of default body map in Chapter 7.

5.3   Where Can One Feel Bodily Sensations?

We have seen that the body map theory can accommodate referred sensations in tools, in phantom limbs, in prostheses, and in multiple primarily concerned with bodily location. The body map indeed represents the structural organization of the various segments of the body independently of current posture. Nonetheless, the body map also contributes to fix egocentric locations. One needs to know the configuration and the metrics of the body in order to know the relative position of all its segments. For instance, when your hand is stretched, the distance between the tip of your fingers depends on their size. Similarly, your arms can be crossed or not with the very same joint angles depending on their size and on the width of your shoulders. Egocentric locations are ultimately grounded in the body map. Consequently, when the body map is altered, not only is the bodily location altered but also the egocentric location. This is well illustrated by the study with the crossed sticks described in Chapter 2: participants have difficulty in judging which stick is stimulated first when the sticks (but not their hands) are crossed (Yamamoto and Kitazawa, 2001b). The difficulty that the participants experience results from the fact that the egocentric location is computed relative to the sticks, which are crossed, rather than relative to the hands, which are uncrossed. In other words, the sticks are represented within the body map and their embodiment has displaced the egocentric location of the sensation. Another example of disruption of egocentric location can be found in patients with auto-topagnosia, whose body map is disturbed. In one study, a patient was asked to construct a pictorial image of the body and of the face with pre-cut puzzle pieces (Guariglia et al., 2002). She produced an odd image of the face with the nose next to two ears on each side next to the eyes. Yet she was happy with the result, demonstrating she was unable to judge the oddity of her work. Interestingly, because of the distortion of their body map, patients with autotopagnosia are unable to correctly point to where they are touched[1]:

When she was requested to show where her elbow was, she looked at her body, searching for the elbow and repeating "yeah elbow, the elbow must be over here", while she was pointing near to the wrist and along the arm.

(Sirigu et al., 1991, p. 631)

---

[1] It is worth noting that autotopagnosic patients do not experience free-floating sensations, they only mislocalize them.

Another consequence of the dependence of egocentric location on the body map is that one cannot experience egocentric location without having information about bodily location (Brewer, 1995). Imagine that I touch your left hand, which is on the left side of your body. If you had experienced only the egocentric location of the tactile sensation, then you should experience the sensation somewhere on the left. Now you cross your hands. You should then experience the sensation somewhere on the right. You should thus experience the sensation moving with no fixed spatial location. As seen in our discussion of the sensorimotor views, this sounds hardly plausible. The reverse dissociation, however, is found in deafferented patients, who can experience bodily location without egocentric location. Such patients localize bodily sensations in their right hand although they can have no idea where their right hand is in the egocentric frame.

To recapitulate, bodily sensations are experienced as being located in a particular part of the body in a particular region of the world thanks to the spatial information conveyed by the bodily senses, but this information does not suffice to fully account for the spatiality of bodily experiences. More is needed for them to acquire their relatively rich spatial content, and more particularly the body map is needed to provide the set of loci with respect to which the location of bodily sensations can be defined.

5.2 The Rubber Band Hypothesis

According to the body map hypothesis, the spatial content of bodily experiences is shaped by the body map, which can be distorted, and includes extraneous objects (for a review, see Holmes and Spence, 2005). But are there limits to the malleability of the body map, and thus to where one can feel bodily sensations? The body map may sometimes seem so plastic that one may wonder what biological constraints, if any, it must conform to. Consider the following example from O'Shaughnessy:

The hypothesis is that if in general one took oneself to be (say) octopus-shaped instead, then despite having a human shape and despite the presence of posture-caused phenomena like sensations of posture, one could not have the experience of seeming to be in the presence of a flexed (very roughly) arm-shaped thing.

(O'Shaughnessy, 1995, p. 184)

If one had an octopus-shaped body map, could one feel that one had tentacles, and if one does, could one feel that one had eight of these? More generally, to what extent can the body map be stretched?

5.2.1 A Liberal Model

Let us first consider the constraint of bodily shape. How anatomically plausible is the body map if it can incorporate non-bodily shaped tools? In order to answer this question, one needs to understand how the various parts of the body are individuated. Contrary to what one might expect, this is far from easy. It is not even clear that the way in which one individuates body parts is universal. Recent cross-cultural studies have shown that linguistic categorization of body parts can vary substantially (Majid et al., 2006). For instance, in some languages, like the Papuan language of Rossel Island, there is just one term for hand and arm, referring to the whole upper limb (Levinson, 2006). This is not an exception to a general rule, as it is true in a wide number of languages (e.g. almost one third of the languages studied in Brown, 2005).

The individuation of body parts can actually appeal to several criteria. It can be based on functional salience (Morrison and Tversky, 2005) or it can follow visual features, including shape, size, and spatial orientation (Andersen, 1978; Brown, 1976). It can be even less fine-grained, using only "geons" (simple 2D or 3D forms, such as cylinders, bricks, wedges, cones, circles, and rectangles) delimited by visual discontinuities (Biederman, 1987). One might then even envisage that a box, which is not that much different from a geon, could be included in the body map, and as we shall see, this is actually possible (Hohwy and Paton, 2010).

Even if one feels that one has octopus arms, can one feel that one has eight of these? According to a *liberal model*, there are some degrees of liberty relative to the anatomical template of the human body, and the body map can represent the body with more than two hands. What seems to argue in favour of such a model is that the experience of a phantom limb can be found not only in amputees, in which the phantom fills the gap of the missing limb, but also in individuals who experience supernumerary phantom limbs despite the fact that they already have two arms and two legs. Interestingly, the supernumerary limbs can feel as real as the biological limbs: "Examiner: But there must be 2 real and 2 unreal? Patient: I guess this is so but they seem all the same, all real to me" (Vuilleumier et al., 1997, p. 1544). A patient even reported feeling her supernumerary hand touching her shoulder, experiencing sensations both in her supernumerary hand and in her shoulder. When she scratched herself with her supernumerary hand, she felt relief from the itching sensation. In addition, she reported seeing her supernumerary hand, which looked "pale", "milk-white", and "transparent" (Khateb et al., 2009, p. 699). Her brain activity in somatosensory, motor, and visual areas during movements of the supernumerary hand confirmed her introspective reports.

The liberal model seems also to be confirmed by a version of the RHI, known as the supernumerary hand illusion. Whereas in the classic version of the RHI, one's own hand remains hidden from sight, in the supernumerary hand illusion, one can see both one's hand and the rubber hand being stroked (Guterstam et al., 2011). What is interesting is that participants report feeling sensations in their hand and in the rubber hand. Furthermore, when asked whether it seems as if they had two right hands, they agree. In another version of the RHI, participants were shown two rubber hands placed side by side above their biological hand, which was hidden behind the table (Ehrsson, 2009; see also Newport et al., 2010; Folegatti et al., 2012). After synchronous stroking of the three hands (i.e. the two rubber hands and the participants' hand), they reported feeling touch on both rubber hands.

5.2.2 A Conservative Model

Do the foregoing studies support the liberal model of the body map? Perhaps, but there are alternative explanations of supernumerary limbs that are compatible with a more conservative model. A first question is whether all the hands (phantom hands, rubber hands, and biological hands) are represented together within the body map. It has been argued that in the classic RHI, the embodiment of the rubber hand is at the cost of the embodiment of the biological hand (for discussion, see Lane et al., 2017). The fact is that participants deny that they experience having three hands (Longo et al., 2008). This seems to be confirmed by the finding that there is a drop in temperature of the biological hand only in the synchronous condition (Moseley et al., 2008) but this finding has been difficult to replicate and the drop in temperature may be found in both synchronous and asynchronous conditions (Rohde et al., 2013). More interestingly, it has been shown a decrease in motor excitability in the biological hand only in the synchronous condition, which may be

THE BODY MAP THEORY 93

hands but only because these sensations are still felt within the boundaries fixed by the body map. But are there sensations whose felt location is not even constrained by the body map (i.e. exosomesthesia)? In other words, can one feel bodily sensations anywhere, maybe as far as the moon, as suggested by Armel and Ramachandran (2003)?

> If you looked through a telescope at the moon and used an optical trick to stroke and touch it in synchrony with your hand, would you "project" the sensations to the moon? (Armel and Ramachandran, 2003, p. 1500)

In more than a century of research on bodily sensations, a few cases of apparent exosomesthesia have been described. If these sensations were truly localized independently of the frame of reference provided by the body map, then they would invalidate the body map theory. But are they real cases of exosomesthesia?[3]

5.3.1 Bodily Illusions in Non-Bodily Shaped Objects

Two studies have recently questioned the importance of the body for experiencing "bodily" sensations. In the first study by Hohwy and Paton (2010), both a rubber hand and the participants' hand are initially stroked, but the rubber hand is suddenly replaced by a small white cardboard box. This does not prevent participants reporting sensations on the cardboard box. The authors claim that this illusion is "explaining away the body", but in what sense? It is worth noting that the experimenter cannot induce the illusion for the box if no prior classic RHI has occurred before. Hence, the body map definitely plays a role. The study only shows that once normally elicited by a rubber hand, visual capture of touch is not disturbed by the intrusion of an object. One possible interpretation of this result is that the transition from the rubber hand to the box is perceived as a visual distortion of the hand (something like "my hand looks like a box"), to which the body map adjusts. If this interpretation is correct, then Hohwy's and Paton's illusion merely shows the flexibility of the body map. Alternatively, as suggested by the authors themselves, it might be that the box is perceived as hiding the hand:

---

[3] This debate is partly orthogonal with respect to Fulkerson's (2014, Chapter 6) discussion of what he calls tactual projection. His objective is to deny that the sense of touch is a contact sense. Instead, he argues, touch involves direct or indirect tactual connection with objects. He further argues that tactual objects are felt to be located in peripersonal space. But he does not address the question of where tactile sensations, instead of tactual objects, are felt to be located and focuses mainly on the exteroceptive side of touch.

participants simply feel sensations to be located in their hand, which they localize in the box. Both interpretations are thus compatible with the body map theory.

The second study that might pose a challenge to the body map theory uses the cutaneous rabbit illusion: repeated rapid tactile stimulation at the wrist, then near the elbow, can create the illusion of touches at intervening locations along the arm, as if a rabbit had hopped along it. In Miyazaki et al.'s (2010) version of this illusion, participants lift up a stick between their two fingers until it is in contact with the system that delivers mechanical pulses on the fingers via the stick. They receive a series of tactile stimulations on their left index finger, then on their right index finger. Participants then report feeling touches between the two fingers, that is, on the stick that they are holding. The authors conclude that tactile sensations can "hop out of the body". Hop out of the biological body, yes, but do they hop out of the body map? This is less certain. Tactile sensations experienced on the stick are no more surprising than sensations experienced on tools. Actually, one might even say that the stick is a kind of tool that the participants manipulate to interact with the stimulating device.

These studies show that the body map has some degree of freedom with respect to bodily resemblance (which we knew already on the basis of the possibility of referred sensations in tools). In this sense, the body can be said to be explained away, but explaining away the body is not explaining away the body map.

5.3.2 Sensations in Remote Objects

A possibly more troubling case can be found in a study by Von Békésy (1959), which seems to indicate that one can feel tactile sensations to be located in external objects not only with no resemblance with the body but also with no spatial contiguity:

> But if the observer was permitted to see the movements of the loudspeaker in the room and coordinate them with the sensations on his arms, after some training he began to project the skin sensations out into the room.
>
> (Von Békésy, 1959, p. 14)

Can one use the same strategy as for the previous findings on referred sensations? In other words, is it plausible that one incorporates the loudspeaker into one's body map? Such an explanation has been tried

(Martin, 1993; Smith, 2002), but I think with little success. This strategy indeed assumes puzzling distortions of the body map,[4] but the even more important difficulty that it faces is the lack of an independent reason to assume that the loudspeaker, rather than any other object in the room, is incorporated into the body map. One cannot appeal to the fact that the observer felt sensations in it due to risk of circularity, and there seems to be no other plausible justification for the claim that the loudspeaker is incorporated. The observer has never interacted with it. Moreover, it cannot be explained as a kind of RHI. The RHI involves *fusion* between a visual event and a tactile event, whereas what the observer might have perceived here is a relationship of *causality* between the movement of the loudspeaker and the sensation. Causality involves a relationship between two distinct events, not a fusion between them.

Since there is no valid reason to claim that the loudspeaker has been incorporated into the body map, one might think that Von Békésy's case casts doubt on the body map theory. However, this is not so for the simple reason that it is only a case of *indirect* perception. To recall, perception is indirect if the information about the perceived object or event is not fully embedded in the information about the more proximal object and further knowledge is required (Dretske, 2006). As Von Békésy acknowledges himself, it was only "after some training" that the observer had learnt to project the sensation. We are thus far from the blind man's direct sensations in his white cane that we saw in Chapter 2. Von Békésy's case distinctly shows that the body map theory does not apply to *indirect* referred sensations, but it says nothing about direct referred sensations.

5.3.3 Sensations in Peripersonal Space

So far I have only considered cases in which one feels sensations in things external to the body, whether they are tools, boxes, sticks, or loudspeakers. Although these entities bear little resemblance to body parts, they are

[4] In order to defend this interpretation, Smith (2002, pp. 136–7) notes that in the Pinocchio illusion one can feel one's nose stretch a long way. So one can conceive that the representation of the body may expand to include a loudspeaker a metre outside the body. Alternatively, he suggests that the representation of the body could be discontinuous, one piece at the actual location of the body and another piece at the location of the loudspeaker. However, it appears that bodily contiguity is crucial for embodiment (Tieri et al., 2015a).

at least like body parts in one respect: they are material *objects*. As such, they could conceivably be represented as parts or extensions of the body. There are, however, cases in which one feels sensations as being located in a specific *empty region of space*. Can the body map theory accommodate such cases?

Let us first reconsider Hohwy and Paton's (2010) study. In one condition they stroke a discrete volume of empty space five centimetres above the rubber hand in synchrony with the biological hand. Interestingly, participants report that they still feel sensations on their own skin, and not above it. A subject, for instance, describes it as follows: "it's a magnetic field impacting on my arm" (p. 8) (see also Guterstam et al., 2016 for a similar illusion). The point here is that even when the stimulations are not on the body, the subjects can still experience them on their body. There is, however, another version of the RHI, called the invisible hand illusion (Guterstam et al., 2013). In this study, the experimenter synchronously strokes the hidden participant's hand and a discrete volume of empty space above the table in direct view of the participant. This time, participants localize their sensations of touch at the empty location. Von Békésy (1967) also reports a similar type of referred sensation. By placing two vibrators slightly out of phase with each other on two spread fingers or on the outspread thighs, healthy subjects describe feeling the vibration in the region of empty space between the fingers or the legs. Finally, similar reports are also found in patient studies. An amputee described feeling a sensation "in space distal to the [phantom] finger-tips" when his stump was stimulated (Cronholm, 1951, p. 190). Another patient "mislocalized the stimulus to the left hand into space near that hand" (Shapiro et al., 1952, p. 484).

How should one interpret these puzzling cases? Are these referred sensations completely disconnected from the body? Can one feel sensations not on the moon itself but simply up in the sky? The reply that I want to offer is negative. It is crucial to note that in all these cases, referred sensations are localized close to the body, in the space also known as *peripersonal space*. Peripersonal space can be conceived of as a buffer zone in which boundaries between the body and external objects are blurred. As Graziano and Gross (1993, p. 107) describe it, it is like "a gelatinous medium surrounding the body that deforms whenever the head rotates or the limbs move". Numerous studies in monkeys and humans, in both healthy and pathological conditions, have explored the functional features of this specific area close to the body (for a review, see Brozzoli et al., 2012).[5] Interestingly, it was after they had found bimodal neurons activated both by tactile stimuli on the skin and by visual stimuli presented in the space *close to the body* of a monkey that Rizzolatti and his colleagues (1981) first coined the term "peripersonal space". In humans, it has been shown that visual stimuli that are presented within peripersonal space can interfere with tactile processing (i.e. cross-modal congruency effect).[6] Such visuo-tactile interference can occur only there because it is only in peripersonal space that both visual and tactile experiences share a common spatial frame of reference centred on body parts (Kennett et al., 2002).

One way to interpret the influence of visual experiences on tactile experiences is to say that the perceptual system anticipates the contact of the seen stimulus on one's body, which allows for enhanced processing of the forthcoming event (Engel et al., 2001; Hyvärinen and Poranen 1974; Kandula et al., 2015; Pia et al., 2015). Specifically, the sight of objects close to one's body can generate expectation of a tactile event, which then influences the experience of the actual tactile stimulus. What is interesting is why such an expectation is generated. One explanation is that the perceptual system expects the body to move. As we shall see in more detail in Chapter 9, peripersonal space is indeed the space "within which it [the body] can act" (Maravita et al., 2003, p. 531): it is the space where the body could be in a near future. Peripersonal space is thus a grey zone between one's body and the external world. Interestingly, it has been repeatedly shown that the RHI works only if the rubber hand is placed in

---

[5] I shall return at length to the notion of peripersonal space in Chapter 9.

[6] In the original study by Spence and colleagues (2004), participants were asked to perform a rapid discrimination task concerning the location of a vibro-tactile stimulus presented either on the left or the right index finger or thumb, while trying to ignore visual distractors presented simultaneously at either congruent or incongruent positions. Crucially, incongruent visual distractors interfered with tactile discrimination (i.e. participants were both slower and less accurate), but only when visual stimuli were close to the body, up to 30 cm. A similar effect can be found in the neuropsychological syndrome of tactile extinction. After right-hemisphere lesions, some patients have no difficulty in processing an isolated tactile stimulus on the left side of their body but when they are simultaneously touched on the right hand, they are no longer aware of the touch on their left hand. Interestingly, the same happens when they see a visual stimulus near their right hand: the visual stimulus on the right side "extinguishes" the tactile stimulus on the left side so that they fail to detect the touch (Di Pellegrino et al., 1997). Similar multisensory effects have been found with auditory stimuli (e.g. Canzoneri et al., 2012).

peripersonal space (Lloyd, 2007; Preston, 2013). Roughly speaking, what is in peripersonal space could be part of one's body. In the invisible hand illusion, the region of space that is stroked is also within the limits of peripersonal space.

We can now see how peripersonal sensations (i.e. bodily sensations felt in peripersonal space) are compatible with the body map theory. When an object or event enters peripersonal space, it is automatically encoded in relation to bodily boundaries as fixed by the body map. Under normal conditions, the perceptual system then generates tactile expectations, which can in turn generate tactile sensations, which are localized on the body. This involves a remapping of what occurs in peripersonal space onto the surface of the body. In illusory or pathological conditions, I suggest that this remapping can be disrupted. In the invisible hand illusion, sensations are still localized within bodily space, but the body is taken to be at a different location from where it is actually. In pathological conditions, the remapping simply fails to occur and sensations remain localized within peripersonal space. Peripersonal sensations are thus only the consequences of the exceptional disruption of the normal process of remapping in tactile expectation.

To conclude, thanks to the malleability of the body map, the localization of bodily sensations is not constrained by the biological limits of the body, but this is not say that there are no constraints. It does have limits, those of the body map. Here I have analysed a series of apparent cases of exosomesthesia, arguing in each case that there is an alternative interpretation that respects our basic intuitions about bodily sensations.

5.4 Conclusion

I have argued in favour of a dual specification of bodily experiences both in terms of bodily senses and in terms of body map. The bodily senses do not fully account for the spatiality of bodily experiences. This is not to say that they carry no spatial information, but only that the spatial information that they carry is limited in scope. It needs to be supplemented by the body map. Bodily experiences acquire a relatively rich and accurate spatial content only when raw spatial somatosensory signals are interpreted through the lens of a topological and geometric map of the body. Bodily experiences are then localized relative to the boundaries of the body as they are represented in the body map.

However, the body map might appear as a *deus ex machina*: it seems to solve all our problems, but its origin is dubious. Furthermore, it is not even clear that it suffices to account for the spatiality of bodily experiences. We have seen that the body map is necessary for egocentric location, but what other information is needed to derive egocentric locations from bodily locations? The discovery of the RHI, along with other recent empirical findings on multisensory interaction, may offer some answers to these questions. We shall see that bodily awareness involves a broad suite of multisensory effects. One should not conceive of these interactions as mere side effects of the way the brain is hard-wired with no bearing on the nature of bodily experiences. Rather, their presence has fundamental implications for bodily awareness.

6

A Multimodal Account of Bodily Experience

Psychology and neuroscience have traditionally addressed each sensory modality in isolation. This is in line with a modular conception of perception, according to which the first levels of sensory processing are encapsulated, isolated from the influence of any other kind of information. In particular, it has been assumed that bodily awareness exclusively results from the bodily senses, with no influence from external senses. However, not only do we simultaneously experience the body through multiple senses, both internal and external, but those senses can also influence each other. Information from one modality then affects information in another modality. Multisensory effects occur very early on in sensory processing and they are often automatic and mandatory. They are so pervasive that it may even seem that a purely unimodal experience (not affected one way or the other by information coming from a different modality) is more the exception than the rule.

Consider some of these multisensory effects. The first that comes to mind is of course the RHI. So far we have considered it from the perspective of the sense of bodily ownership, but the RHI also illustrates how vision can alter the felt location of one's hand. Another well-known visuo-tactile effect is known as visual enhancement of touch: spatial acuity increases if one sees the touched body part, although one does not see the object that is touching it (Kennett et al., 2001; Press et al., 2004; Tipper et al., 1998). Other bodily illusions highlight the role of auditory information for bodily experiences. In the parchment skin illusion, one reports feeling that one's skin is dry, parchment-like after hearing a distorted recording of the sounds produced by one's hands rubbing back and forth (with high frequencies accentuated) (Jousmaki and Hari, 1998). Consider also the marble-hand illusion: one reports feeling one's hand as stiffer, heavier, and harder after hearing the sound of a hammer against the skin progressively replaced by the sound of a hammer hitting a piece of marble (Senna et al., 2014). The interaction between bodily experiences and visual or auditory information is such that the mere presence of visual/auditory information can induce bodily experiences. We are all familiar with the unpleasant bodily sensation triggered by the sound of fingernails scratching a chalkboard or with the sudden sensation of itching when seeing someone scratch her leg.

Multisensory effects are thus pervasive in bodily awareness and part of my objective in this chapter will be to review them, and in particular to analyse the many ways vision impacts on proprioception and touch. Indeed multisensory effects do not form a homogeneous category. Distinctions need to be made between effects that involve recoding one modality into the format of another modality (i.e. conversion) and those that involve recoding in a common amodal format (i.e. convergence), between effects that are fleeting and short-term and those that are stable and durable, and between effects that are perceptual, attentional, and cognitive. In the face of them, it seems difficult to defend the view that bodily experiences can be reduced to bodily senses. Instead, I shall argue that bodily experiences are constitutively multimodal (Vignemont, 2014a).[1]

6.1 The Visual Body

In Chapter 5, I highlighted the limited scope of information that is carried by bodily senses. But not only are their signals relatively poor on their own but they are also of limited reliability as far as spatiality is concerned (Rossetti et al., 1994; van Beers et al., 1998, 1999; Helms Tillery et al., 1991). We all know how clumsy we are when moving around in the dark. We might believe that it is because we do not see our environment but even if we do see the target that we are trying to reach (but not our arm), our performance is still relatively poor (Cameron et al., 2015). Proprioception indeed does not suffice for reliable correspondence between the location of the body in the external world and bodily experiences.

---

[1] I use the term "multisensory" to refer to the backstage mechanisms at the subpersonal level, and the term "multimodal" for the personal level. Further useful distinctions may be found in Macpherson (2011) and Spence and Bayne (2014).

Other sources of information are required, both to supplement for informational limits (such as information about the length of the segments connecting the joints and the width of each body part) and to improve reliability. Vision is then a good candidate to fulfil both functions because it provides highly accurate and rich spatial information, which is usually more reliable information than touch and proprioception.

6.1.1 Visual Egocentric Location

When you lie completely motionless in bed at night, you may sometimes feel uncertain of the exact location of your limbs. Proprioceptive signals are weak when one is not moving and, after a few minutes of complete stillness, one can lose one's body, so to speak. In order to be as reliable as possible about the location of one's limbs, one must move, but even then the accuracy of proprioception is limited and it decreases with the number of joint angles that must be computed: the more distal the body part, the more complex the computations. For example, information about the location of one's fingertip in space requires taking into account information from many receptors on many joints, muscles, and tendons, each of them sending noisy signals, the noise increasing in proportion to the degree to which the body part is distal. As Longo and Haggard (2010, p. 658) summarize the situation: "There is . . . no afferent [somatosensory] signal, or combination of afferent signals, analogous to a global positioning system (GPS) signal".

It is thus optimal to combine visual and proprioceptive information in order to achieve the most reliable perceptual estimate of egocentric location (van Beers et al., 1999). Hence, one first relies on visual information, which dominates the other sources of information.[2] Even when visual information is manipulated so that it is systematically erroneous, it is only after some time that one stops trusting vision and eventually relies more on proprioception (Bellan et al., 2015). The major contribution of vision to bodily experiences is well illustrated by prismatic adaptation (i.e. prisms make participants see their hand at a location distinct from where it actually is). After a while, participants

---

[2] However, in some perceptual contexts, proprioception is more accurate than vision, for instance when the hand is actively moving and for certain spatial directions (e.g. depth). In these cases, more weight is given to proprioception than to vision in the integrative process (Van Beers et al., 2002).

wearing prisms cannot help but *feel* their limbs at their visually perceived position and they can no longer retrieve their "pure" proprioceptive location independently of the influence of the prisms (Stratton, 1899; Welch and Warren, 1980). As described by Stratton (1899), there is then a "harmony of touch and sight".

The involvement of visual information in bodily experiences is actually more pervasive than one might expect. Let us consider again the RHI. When participants mislocalize their hand towards the location of the rubber hand, their bodily experiences take into account visual information about the location of the rubber hand, although they now have no current visual experience of either hand. The weight given to the visual information is such that it has been found that participants automatically make small compensatory movements to reduce the conflict between proprioception and vision (Asai, 2015). Another study shows that even if after the stroking they have moved their hand to grasp a stick in front of them and then returned it to the same location, they still verbally mislocalize their hand (Kammers et al., 2009a). This shows that the influence of vision extends even when online visual input is withdrawn and proprioceptive signals updated.

It has also been found that bodily experiences can result from the integration of current proprioceptive information and visual *prediction* of hand location (Smeets et al., 2006). In one study, participants are asked to move a cube with and without visual feedback from their hand movements. The analysis of the drift of the hand permits evaluating to what extent the visual estimate is taken into account. When the light is turned off, participants still locate their hand where they have last seen it, even when the current position differs. More surprisingly, when they start moving in the dark, they still use the visual estimate of their hand location, which is updated on the basis of their intention to move. It is only after several movements that participants rely less on their visual estimate because each movement of the unseen hand adds uncertainty to their visual estimate. The results thus show that when one moves one's hand, one anticipates the sensory consequences of one's movements, including the expected visual location of the hand. Furthermore, they show that the bodily experience that represents the location of the hand can be updated on the basis of visual expectation, even when one is in the dark. Hence, the involvement of visual information is not restricted to cases in which one actually sees one's body.

6.1.2 The Synaesthetic Body Map

That our access to the external world, and to the position of our body in it, is partly based on vision may not seem surprising. But our access to the intrinsic properties of our body is based on it too. O'Shaughnessy (1989, p. 56) claims, "the body image must be the image of what would reveal itself in tactile encounters". He thus seems to suggest that the body map is eventually nothing more than a collection of information from bodily senses over time. However, such a tactile conception of the body map seems counterintuitive in light of the limits of touch, and more generally of bodily senses. It is true that the somatosensory feedback that one receives when acting carries information about the size of the limbs, but this feedback provides only a rough estimate. In particular, to know how far one can reach with one's hand does not indicate the respective size of one's fingers, palm, forearm, and upper arm. Active exploration of each body part by haptic touch seems to fare better and to be more specific. However, this involves complex tactile-proprioceptive processing, and that in turn requires taking into account the size of the exploratory body parts (e.g. fingers).

In addition, O'Shaughnessy seems to neglect that our acquaintance with space is primarily grounded in vision and bodily space is no exception. The RHI is a vivid illustration of the role of vision not only for egocentric location but also for bodily location. Indeed, the body map has adjusted to incorporate a rubber hand that was only seen, and only for a couple of minutes. This seems to indicate that it can be quickly updated on the basis of visual information. Other studies have found that visually modifying the size of your limbs can affect your tactile experiences. For instance, if you are touched on two spots on your finger after observing the image of a larger version of your hand, you experience the distance between the two tactile stimuli as relatively larger (Taylor-Clarke et al., 2004): the visual distortion alters the body map, which in turn calibrates tactile processing. Even more surprisingly, viewing one's hand actively using a tool, although it is actually stationary, can induce the same effect, thus showing that visual inputs about tool use can suffice to modify the body map (Miller et al., 2017). The impact of vision can be explained by the fact that vision is essential to fill in the metric details of the body map. It is indeed the only sensory modality that can directly and reliably (though not perfectly) process size information (Longo and Haggard, 2010).

Vision thus participates in the building up of the body map, vision of one's body, but possibly also vision of other people's bodies. Consider the case of amelic patients who are born without legs or arms. As mentioned in the Introduction, they can experience phantom limbs. One can explain their phantom limbs if the body map innately encodes a rough specification of a human body (such as two arms and two legs), regardless of the physical body that they were born with (Melzack et al., 1997). However, the innate body template does not suffice to provide the fine-grained details of their phantom limbs.[3] Another factor that may contribute is the constant visual awareness that amelic patients have of other bodies with two arms and two legs. On its basis, they can enrich their own body map (Brugger et al., 2000; Price, 2006). Vision of other bodies can also play a role in shaping the body map in healthy individuals. There are indeed parts of the body that are not directly visually available, such as the back. But even if we cannot see our own back, we see other people's backs and visual information about their bodies may supplement the representation of the configuration of our own body. In Chapter 7 we will actually see that some components of the body map are interpersonal.

To recapitulate, visual information about one's own (and possibly other people's) body parts shapes the body map that spatially structures bodily experiences. The body map should thus be conceived of as a unified multimodal representation that results from the integration of innate, somatosensory, and visual information (e.g. Mancini et al., 2011). As concluded by the neurologist Schilder a long time before multisensory research had started to take off:

It is not the case that the schema of the body has two different parts, the one optic and the other tactile. It is in essence a synaesthesia (. . .) The synaesthesia, therefore, is the normal situation. (Schilder, 1935, p. 38)

6.2 The Multimodality Thesis

It is now time to assess the implications of these results for bodily awareness. Do they show that vision plays only a *causal* role in fixing the spatial content of bodily experiences? Or do they show that bodily

---

[3] For discussion of innate body representations, see Bremner (2017).

experiences are *constitutively* multimodal? The notion of multimodality has become of interest to philosophers only recently (O'Callaghan, 2011; Macpherson, 2011; Spence and Bayne, 2015; Bennett and Hill, 2014). Unfortunately, there is already little agreement in philosophy on the way to individuate the modalities themselves, let alone on how to understand multimodality. If by multimodality one means the mere *combination* of perceptual experiences in different sensory modalities, then it is hardly controversial that some bodily experiences are multimodal. The body is what Aristotle called a "common sensible", which is accessible through multiple senses. When I am aware of my body, I proprioceptively and interoceptively feel it, I touch it, I see it, I can also hear it. I do not have all those experiences independently of each other. Instead, they are bound together into a unified multimodal experience of my body. However, in light of the effects described in the first section, we might want to go further in our analysis. Here I will defend what I call the multimodality thesis, according to which bodily experiences are *constitutively* multimodal in virtue of their specific aetiology, which has been selected for veridically representing bodily properties.

6.2.1 The Binding Model of Multimodality

Let us try to refine our definition of multimodality. Multimodal unity can be compared to the unity of visual experiences that represent several features of the same object (e.g. the colour and the shape of the book that I see are unified into a visual experience in virtue of being the colour and shape of the same object). Like visual unity, multimodality unity involves *perceptual binding*. There is perceptual binding if different pieces of information taken to be about the same object or event are brought together into a unified content. Binding is successful if the information that is bound is actually about the same object or event.

There are two forms of multisensory binding: additive binding and integrative binding (Vignemont, 2014b). In *additive binding*, modality-specific experiences are combined and they complement each other. Imagine that I feel my arms crossed and simultaneously see that they are tanned. Thanks to additive binding, I have a multimodal experience of my arms. Additive binding results from the combination of two types of sensory information that are not redundant because they are about distinct features or distinct parts of the same object. The perceptual system can then collect all the pieces of information about the various parts and the various features of an object in order to have as rich and complete an experience of the object as possible. *Integrative binding* is more narrow-minded, so to speak. It focuses on a single property of the object. It results from the fusion of distinct types of sensory information that are redundant. For instance, I feel through proprioception that my arms are crossed and I see them crossed. The proprioceptive experience and the visual experience, which are both about my bodily posture, are merged together into the unified content of the multimodal experience. Because of the redundancy, the binding can be so strong that the experiences can blend into each other, so to speak. Whereas additive binding aims at completeness, integrative binding aims at reliability. Each sensory receptor sends noisy signals. Furthermore, the quality of the signals can be decreased by environmental conditions (e.g. poor light or noisy environment). It is thus important to have more than one source of information. Informational redundancy increases robustness and reliability. Thanks to integrative binding, the perceptual system can generate the best estimate of the property of the object by integrating various sources of information. From now on, I will exclusively focus on integrative binding.

6.2.2 A Constitutive Thesis

In light of the examples of multisensory interactions described so far, it may be said that some bodily experiences are multimodal thanks to their specific aetiology, which involves multisensory integrative binding (the RHI, for example). This fact, which has often been neglected, is interesting in its own right. However, I want to make an even stronger claim and argue that bodily experiences are *constitutively* multimodal. More specifically, I will argue that multisensory integrative binding is a constitutive component of the aetiology of bodily experiences.

Certain types of causal relations and dependencies can amount to constitutive relations when they have been selected to contribute to the fulfilment of a given function. Let us assume here that the function of bodily experiences is to afford a veridical rendering of bodily properties (O'Shaughnessy, 1980, vol. 1). In other words, bodily experiences have been selected to reliably track bodily states. The function of bodily experiences is fulfilled if and only if bodily experiences are reliably correlated with bodily states that they are designed to indicate. Let us further assume that the aetiology of bodily experiences is partly determined by their function. The question has now become whether vision is required by the aetiology of bodily experiences for them to achieve their function of veridically representing bodily properties. In my view, the answer is positive.

The contribution of vision to bodily experiences is highly specific, compared to other factors that can influence bodily experience. For example, our mood can influence how we perceive our body but it is unclear how this helps in reliably tracking bodily states. By contrast, multisensory binding is required in order for bodily experiences to fully realize their function. Visual information, which is characterized by its spatial richness and reliability, is needed to compensate for the flaws of bodily senses. As Stein and Meredith claim (1993, p. 6), "the sensory modalities have evolved to work in concert". This applies to vision and bodily senses. Their interaction improves the likelihood of detecting, localizing, and identifying bodily events and properties. It is thus beneficial to combine different sources of information in order to achieve the best bodily judgements. The more information, the better. Hence, the mechanisms of multisensory binding have been selected by evolutionary pressure because they are required by the very nature of bodily experiences. Consequently, the contribution of vision to bodily experiences is not a mere causal accident. In this sense, bodily awareness is constitutively visual.

The multimodality thesis, however, seems to immediately face the following objection. The contribution of vision to bodily awareness is not of the same type as the contribution of bodily senses. For example, if one receives no information from the bodily senses, then one experiences no bodily sensation, as shown by peripheral deafferentation. Some deafferented patients may even report that they feel as though they are "nothing but a head" (Gallagher and Cole, 1995). By contrast, the loss of vision does not lead to such an extinction of bodily awareness. One might conclude that vision cannot play a constitutive role for bodily experiences. However, I do not defend a strong constitutive thesis for multimodality, according to which one could not have bodily experiences if one were blind. This view is obviously false. Instead, I suggest drawing a distinction between a strong and a weak version of constitutive explanations and I defend only a weak constitutive thesis of multimodality, according to which one would experience one's body *differently* if one were blind.[4] I will now show that there are indeed fundamental differences between bodily experiences in those who see and in those who have never seen: (i) blind individuals are not sensitive to the same bodily illusions; (ii) they do not use the same strategy to compensate for the flaws of somatosensory information, and (iii) they do not exploit the same spatial frames of reference. Without vision, bodily experiences cannot fulfil their function to the same extent and in the same way.

6.3 Losing Sight of the Body

How do blind individuals experience their bodies? They feel touch, pain, and other itches in the various parts of their body. So in what sense are their sensations different from those experienced by individuals who can see their bodies? Let us start with the somatic version of the RHI, which does not appeal to the vision of the rubber hand, participants being blindfolded the whole time: participants stroke a right rubber hand with their left hand, while they are stroked in synchrony on their right hand (see Appendix 1 for more detail). In the original study, participants reported feeling that they were touching their own right hand. When asked to point to their right index finger after the illusion, they mislocated it in the direction of the location of the index finger of the rubber hand (Ehrsson et al., 2005). Since there is no visual component in the illusion, one might expect no difference between sighted and blind participants. Indeed the rubber hand is never visually available and the illusion involves the integration of only tactile, proprioceptive, and efferent information. Yet blind participants did not report feeling as if they were touching their own hand and some even claimed that it was "absurd" (Petkova et al., 2012).[5] When asked to point to their hand, they showed no proprioceptive drift. As the authors concluded: "This finding

---

[4] Another possible example of a weakly constitutive thesis is flavour perception. Gustatory experiences are strongly influenced by olfaction, and it does not seem that olfaction is just one among many other factors that can influence flavour (Auvray and Spence, 2008). Yet we retain some gustatory experiences even during the worst cold, although food tastes differently. On my view, olfaction is constitutive of flavour perception, but only in the weak sense that I have described.

[5] This study did not analyse separately congenitally blind and late blind participants. However, another study by Nava et al. (2014) showed that blind participants remained insensitive to the ownership illusion (as reported in a questionnaire) no matter when they lost vision.

suggests the existence of fundamental differences in central body representation between blind and sighted individuals" (Petkova et al., 2012, p. 8). What are those fundamental differences?[6]

Firstly, individuals who are congenitally blind or have become blind in childhood have a partially distorted body map. The tactile and kinaesthetic information that they receive about their body metrics and configuration cannot fully compensate for the lack of visual information. Without visual input, the perceived size of the various segments of the body is more biased than it is in sighted individuals. Helders (1986) reported that blind individuals represented their torso as long and very narrow with disproportionately big arms and hands. Kinsbourne and Lempert (1980) showed that blind individuals had a less accurate representation of the size of their body parts compared to sighted individuals. As they concluded:

When vision is unavailable for the construction of a geometrically accurate internalised space, the other spatial senses construct a body scheme which in general has topological validity but falls short of affording a veridical rendering of the properties of the human body.   (Kinsbourne and Lempert, 1980, p. 37)

The second interesting difference between sighted and blind individuals is revealed in their performance in localizing their body parts and their tactile sensations. Jones (1972) compared two conditions: participants (sighted and congenitally blind) could have their eyes open or closed. In the eyes-open condition, congenitally blind subjects were less accurate than sighted subjects, but in the eyes-closed condition, they outperformed them.[7] These findings indicate that the difference between sighted and blind bodily experiences cannot be reduced to the presence versus absence of online visual input because even with eyes closed, the performance of blind subjects differs from that of sighted subjects. There are two reasons for this. On the one hand, the processing of bodily

---

[6] Most studies compare haptic perception of objects in blind and sighted individuals, and only a few of them directly investigate bodily experiences in blind people. In the studies that I describe, the blind subjects had no residual vision: it made no difference for them if their eyes were closed or open.

[7] Further studies found that tactile acuity was better in both early and late blind participants in studies in which sighted and blind participants had their eyes closed (e.g. Goldreich and Kanics, 2003; Alary et al., 2009; Yoshimura et al., 2010).

information in those who can see is shaped by the almost constant involvement of visual information even in the absence of online vision through visual memory and visual prediction. The side effect of this otherwise advantageous feature is that sighted individuals fail to utilize somatosensory information for bodily experiences when they should. On the other hand, blind individuals compensate for the lack of vision by dedicating more resources to somatosensory processing, which can improve the accuracy of their bodily experiences (Yoshimura et al., 2010; Goldreich and Kanics, 2003). This is revealed by large-scale cortical reorganization (for a review, see Merabet and Pascual-Leone, 2010). For instance, extensive practice of reading Braille leads to an enlargement of the finger areas in the somatosensory cortex (Sterr et al., 2003). Furthermore, neurons that would normally respond to visual stimulation can be recruited by bodily senses when visual input is entirely absent. Finally, somatosensory inputs to multisensory areas that are recruited for spatial perception and attention increase thanks to the lack of competing visual inputs. For instance, congenitally blind individuals allocate more attentional resources to potentially threatening stimuli and show hypersensitivity to noxious thermal stimulations (Slimani et al., 2014).

We have thus good evidence that sensory processing of bodily properties is distinct between blind and sighted individuals. However, the most convincing evidence that the resulting bodily experiences are of two distinct kinds involves nothing more than the simple manipulation of crossing one's hands. As already discussed in the previous chapters, one takes longer and is less accurate in judging which hand has been touched first when the hands are crossed than when they are uncrossed (Yamamoto and Kitazawa, 2001a). The effect can be explained by the conflict between bodily location and egocentric location of bodily experiences. I shall now argue that the egocentric frame is primarily visual. One can make two predictions on the basis of this hypothesis. If the egocentric frame is visually grounded, then one expects the effect to diminish or disappear (i) when the hands are crossed in a space that is not visually accessible and (ii) when one is blind. Both predictions have been empirically confirmed. Sighted subjects have less difficulty judging which hand has been touched first when the hands are crossed behind their back than in front of them (Kobor et al., 2006). Furthermore, congenitally blind adults do not show any difficulty when their hands are crossed in front of them (Röder et al., 2004).

What is also interesting is that a similar lack of effect has been found in young sighted children, whereas sighted adults who have their eyes closed and non-congenitally blind adults do have difficulty. The explanation is that the visually grounded egocentric frame is built up during development on the basis of visual experiences (Pagel et al., 2009; Bremner and Spence, 2017). In other words, spatial remapping requires past experiences in the visual modality in order to learn its specific spatial frame of reference. Once the format is learnt, one can translate or recode the format of one modality into that of another modality. For instance, location within body-centred frame of reference used by touch can be remapped into the eye-centred frame used by vision. Once set up, the spatial remapping into the visually grounded frame is an automatic low-level sensory process. Spatial remapping solves what can be called the Tower of Babel problem of multisensory binding. Each sensory modality is encoded in its own spatial frame (e.g. eye-centred vision, head-centred audition, skin-centred touch). In order for the information coming from different modalities to be integrated it must be translated into a common reference frame. This common frame can be amodal or the frame of one modality, like vision in our case. The visual frame of reference is actually of special interest because the world with which one interacts is mainly given through vision.

Hence, perceptual experiences in one modality during development can have long-term consequences on perceptual experiences in a different modality. As a result of spatial remapping, one cannot help but take into account the two spatial frames, even when they come into conflict. It is in congenitally blind people and in young children that the remapping does not occur: they have no difficulty in judging where they were touched first when their hands are crossed. This is not to say that congenitally blind individuals and young children have no egocentric reference frame in general. It only shows that in their case tactile information is not automatically remapped in eye-centred coordinates because of the lack of visual experiences at a crucial stage in development. The lack of remapping into the visually grounded egocentric frame also explains why blind individuals do not experience the non-visual version of the RHI. To relate what one feels on one's hand with the rubber hand that one touches, one needs to remap one's tactile sensations in the external space in which the rubber hand is located. Without remapping, the space of tactile sensations and the space of the rubber hand remain distinct. Hence, blind individuals cannot feel touch on the rubber hand. The format of the experience when a sighted individual or a late blind individual is touched is thus different from the format of the congenitally blind individual's experience.

To summarize, the series of findings that I have reviewed here argues in favour of the multimodality thesis. Processing of bodily information has been selected to guarantee that bodily experiences correspond in some reliable way to the body. Both sighted and blind individuals have access to somatosensory information, but somatosensory information has some limitations, which have to be compensated for. The compensatory strategy is then of two different kinds, for those who see and for those who do not see. Sighted individuals use visual information because of its spatial precision, even in the absence of online visual inputs. Blind individuals do not have that option. Instead, they exploit at best what is available to them, namely somatosensory information to which they dedicate more resources. However, this does not suffice and their body map is still partially distorted. Their bodily experiences are also in a different format, which prevents blind individuals experiencing the same difficulty in temporal order judgement and the same bodily illusions. In particular, multimodal bodily experiences enable one to remap what happens on the body into the external world, remapping that does not occur in congenitally blind people. Arguably, the multimodal strategy has been selected by evolution as guaranteeing the most reliable correspondence to the body in individuals endowed with vision. Without multimodality, bodily awareness would not be the same.

6.4 The More Multimodality, the Less Immunity?

We have seen that the role of vision in bodily experiences is not restricted to a couple of bodily illusions. Rather, bodily illusions reveal a fundamental fact about bodily awareness, namely its constitutive multimodality. One may, however, wonder whether this is compatible with another fundamental fact concerning bodily awareness, the fact that it holds privileged relationship to one's own body. We saw in Chapter 3 that bodily experiences can ground judgements that are immune to error through misidentification relative to the first-person: when I feel my arms crossed, I can rationally doubt whether they are crossed or not, but

I cannot rationally doubt that they are mine. Bodily IEM is explained by the special way of gaining knowledge about the body that is afforded by the bodily senses, which ensures that one's own body is the relevant body for the evaluation of the bodily property. One can thus dispense with self-identification of the type "the arms that I feel crossed are mine". Now what happens if bodily experiences are grounded not only in bodily senses but also in vision? We have seen that from some specific visuo-spatial perspectives, the only body that one can see is one's own. As such, vision can guarantee bodily IEM and its involvement in bodily experiences can only reinforce their special link to the body. However, this type of visual experience is more the exception than the rule and our hands can easily be visually confounded, if we play piano four hands for example. Hence, most of the time I need to identify whose body I see and I can rationally doubt that the arms that I see crossed are mine. In that case, vision does not guarantee bodily IEM. Does multimodality then come at the cost of bodily IEM?

Arguably, the mere presence of a single perceptual ground that guarantees IEM (e.g. bodily senses) suffices for securing the IEM of multimodal experiences, regardless of the presence of other grounds. However, this is true only if the interaction between the grounds that results in multimodal experiences does not involve self-identification. Consequently, in order to assess the IEM of multimodal experiences, one needs to understand the basic principles underlying perceptual multimodality, and more specifically, to decide whether they require identifying the body that we see with the body that we feel.

6.4.1 The Multimodal Binding Parameter

As argued, bodily experiences are multimodal in virtue of their aetiology, which involves multisensory integrative binding. Multisensory integrative binding in turn involves four main processes. The first process aims at *selecting* the relevant information to bind in order to avoid bringing together features or experiences that have nothing to do with each other (for example, the seen colour of your hand and the felt location of my hand). This is known as the "parsing problem" in visual binding (Treisman, 1999), or the "assignment problem" in multisensory binding (Pouget et al., 2002). The function of the second process, which is specific to multisensory binding, is to *recode* the information coming from different sensory modalities into the same format in order to solve the Tower of

Babel problem. Indeed the brain cannot simply combine the converging sensory inputs because of differences in format and reference frames between the sensory modalities. The third process aims at solving what can be called the *weighting* problem. The redundancy of information that characterizes multisensory integration does not imply that all information is equally trustworthy and the reliability of each sensory modality varies widely according to the context and to the type of information. The perceptual system thus needs to temper the importance of each modality given the context. Therefore, in case of conflict, it can achieve the most reliable estimate. The function of the fourth process is to *tag* the selected information as coming from the same object or event so that it can be integrated into a unified content. One key question then is what neural mechanism unifies the selected information. Another key question concerns the strength of multisensory fusion. The content is fully unified if one cannot normally retrieve the original information derived from each sensory modality. For example, one cannot experience the colour of an object without experiencing its shape, except in some rare neurological disorders. However, there is a cost to complete fusion in multisensory integrative binding. Imagine that you wear prisms and after adaptation you see and feel your hand at the same location. Because of the complete fusion, the perceptual system can no longer evaluate the importance of the original conflict between vision and proprioception. It is not important if the conflict is small, but if it is large, it is then better to have only partial fusion.

Arguably, if identification occurs, then it is probably at the first stage, that is, at the stage of the selection of the information to be bound together. One can then raise two distinct questions: (i) what mechanism is used to select the relevant information to the exclusion of other information?; (ii) on what basis does this mechanism operate? Attention is generally the answer to the first question (Treisman, 1998), but some results seem to indicate that cross-modal integration can precede cross-modal attention (see Driver and Spence, 1998, for instance). Although interesting, I shall leave aside this debate and I will focus only on the second question about the criterion that the perceptual system uses to select the relevant information, which John Campbell calls the "binding parameter":

By the "binding parameter", I mean the characteristic of the object that the visual system treats as distinctive of that object, and uses in binding together features as features of the same thing.   (Campbell, 2002, p. 37)

The notion of binding parameter, which has been introduced to characterize visual experiences, can be easily applied to multimodal experiences. The *multisensory binding parameter* is the characteristic of the object that the various sensory modalities treat as distinctive of that object, and use in combining or integrating together information as information about the same object or event. The application of the notion of binding parameter to the multimodal domain adds a new constraint. For the multisensory binding parameter to play the role of guiding the selection of the information across several sensory modalities that will be integrated together, it must be a common sensible, which can be either a characteristic available to all the senses, or more liberally, a property accessible to the sensory modalities that are concerned, and only to those.

What are the candidates for the multisensory binding parameter? Whereas many studies on visual binding have used multiple visual stimuli to test how the right colour is bound to the right shape for example, the majority of studies in the multimodal literature have focused on relatively simple situations in which only a single stimulus is presented in each sensory modality at any given time. Under such conditions, the parsing problem has already been solved since there is no question about which information should be bound. Still, on the basis of the literature on visual binding and attention, one can suggest the following two candidates for the multisensory binding parameter: location and object.

Traditional models of attention characterize attention in spatial terms: one pays attention to stimuli at a specific location to the exclusion of stimuli at other locations (Posner et al, 1980). In contrast, more recent models emphasize our ability to track over time discrete objects, instead of paying attention to spatial regions of the visual field (Scholl, 2001). For example, one is able to selectively look at one of two spatially superimposed movies by ignoring the other despite the fact that they share the same locations (Neisser and Becklen, 1975). Furthermore, when subjects are presented with two overlapping objects (such as a box and a line), they are more accurate in judging two properties of the same object than they are of separate objects (Duncan, 1984). Thus, discrete objects can serve as units of attention. In other words, one is able to attend to clusters of features parsed as independent individuals rather than spatial areas. The dissociation between the two types of attention is confirmed in the neuropsychological literature. For example, patients suffering from neglect represent only the right half of a scene presented and leave out figures on the left side, but they can also sometimes represent only the right side of objects, omitting their left side, no matter where they are located in the scene (Driver and Halligan, 1991; Marshall and Halligan, 1994).

Without necessarily adopting the attentional model of binding, we can keep the distinction and claim that the selection of the relevant elements to bind is based either on a specific region of space independently of the objects that it contains or on the object itself. In the latter case, one can suggest that we use sortal concepts to parse the information. Sortal concepts, such as book, hand, glass, and so forth, provide principles of individuation and numerical identity (Wiggins, 2001): they allow one to count the entities and to reply to the question "what is it?". We can thus contrast what I call the spatial hypothesis and the sortal hypothesis of the multisensory binding parameter.

Spatial hypothesis: The perceptual system binds together the sensory states that carry information about the same location.

Sortal hypothesis: The perceptual system binds together the sensory states that carry information about the same object. The sameness of the object is determined through its conceptual identification.

I will now show that the spatial hypothesis is not satisfactory. Does that imply that the sortal hypothesis is true? If it were, then multimodal bodily experiences could not ground judgements that were immune to error through misidentification since they would involve a prior step of identification. But I will argue that the sortal hypothesis is not satisfactory either. I shall then offer another version of the object-based model, which does not involve conceptual identification.

6.4.2 A Space-Based Model

According to the dominant view, the perceptual system binds together the sensory states that carry information about the same location (for visual binding, see Treisman, 1998; Campbell, 2002). This model is consistent with results on the RHI that show that the rubber hand must be close to the location of the real hand for the illusion to occur (Lloyd, 2007; Preston, 2013). Binding then does not involve identification, and thus allows multimodal judgements to be immune to error. Yet location must be ruled out. I will offer here only two counterexamples, the first involving two objects for one location, and the second involving one object for two locations.[8]

How can amputees experience the presence of phantom limbs despite the fact that they can constantly see that the limb is missing and that there are other objects at the location at which they feel their phantom to be? This seems all the more surprising given the extensive literature on the dominance of vision over other modalities: vision should chase away the phantom, but it does not. Why is this so? One can suggest the following explanation: the perceptual system does not confront visual information about actual objects with proprioceptive information about the phantom hand despite the fact that they are experienced at the same location; there is no conflict, and thus no attempt to solve the conflict by cancelling the phantom sensation.[9] In short, integrative binding does not occur despite the similarity of location.

Further findings on phantom limbs confirm that something more than location is required for integrative binding. As described in the Introduction, Ramachandran and Rogers-Ramachandran (1996) invented a virtual reality box that provided patients with the missing visual feedback of the phantom hand. A mirror was placed vertically so that the mirror reflection of the intact hand was "superimposed" on the felt position of the phantom hand; when the normal hand moved, the mirror reflected a moving contralateral hand. This experimental device could induce the illusion of phantom movements even when the phantom hand was felt earlier as being paralysed. Hence, visual information and proprioceptive information became integrated into a unified bodily experience, but only

---

[8] For further criticisms of the spatial criterion for visual binding, see Matthen (2006), Raftopoulos and Muller (2006), and Raftopoulos (2009). It is also worth noting that even advocates of space-based parsing often agree that location is not a sufficient criterion (Treisman, 1999; Campbell, 2006).

[9] One might want to explain the lack of integrative binding by a failure to recode the two sensory modalities into a common spatial frame of reference. This could go along the following lines: the location of the actual objects is represented within an eye-centred frame only, whereas the location of the hand is represented within a body-centred frame only. However, this explanation is not fully satisfying. There are indeed some reports that indicate that the phantom hand is located in a visual frame of reference. For example, Ramachandran and Blakeslee (1998) describe the case of a patient who felt that he was grasping a mug that he was seeing and who began to scream when the examiner moved the mug away from him. The patient's reaction shows that the phantom hand was experienced in the same space as the mug. Hence, it seems that it cannot be only a problem of recoding.

when they concerned the same perceptual object, namely the hand. Phantom sensations show that location is not a sufficient binding parameter and that the perceptual system needs to appeal to further information. Location is not a necessary condition either, as shown this time by prism adaptation. Despite spatial discrepancy, visual information and proprioceptive information about the location of one's hand are indeed fused together. Consequently, if the identity of the object is the same (one's hand), then there is integrative binding, regardless of spatial differences. The parsing of information can be exclusively object-based.

6.4.3 The Sortal Hypothesis

One may then suggest that one uses a sortal concept to "delineate" the boundaries of the object that is the common cause of the various perceptual experiences.[10] For example, the visual information and the auditory information of an approaching car are integrated in virtue of the fact that one recognizes that they both carry information about *a car*. This is not to say that the sortal concept suffices as a binding parameter because one can simultaneously perceive several objects of the same sort, for example several cars in the street. It is not beneficial to bind together properties that belong to different objects, even if they can be grasped by the same concept. The view is rather that one uses sortal concepts to delineate objects *in a specific region of space*.

It is clear that the application of a sortal concept may give expectations about specific perceptual experiences in different modalities and their co-occurrence. Furthermore, the application of sortal concepts has the advantage of neatly answering the common sensible constraint. Arguably, sortal concepts are amodal, and thus common to all sensory modalities. The sortal hypothesis seems also to account for the findings on phantom limbs. However, one may wonder whether the parsing of information requires such a sophisticated explanation, especially since it has the unwelcome consequence that multimodal bodily experiences involve an identification component, and thus can lead to judgements that are sensitive to error through misidentification. Does one need to recognize

---

[10] The sortal hypothesis may look similar to the delineation hypothesis discussed by Campbell (2002). They both highlight the role of the sortal concept, but not at the same level. The sortal hypothesis concerns binding whereas the delineation hypothesis concerns demonstrative reference and conscious attention.

the object that is perceived in order to register that the two sources of information are emanating from a single object?

The sortal hypothesis seems hardly plausible in the light of the neurophysiological data (that show that low-level neural mechanisms are involved in multisensory binding). One does not even need to be conscious of the information that is integrated (Faivre et al., 2015). In one study it has been shown that visual stimuli rendered invisible through continuous flash suppression improve tactile processing and enhance the perception of near-threshold tactile stimuli (Salomon et al., 2017). In addition, participants experience ownership for a virtual avatar after receiving tactile stimuli on their own body and seeing visual stimuli on the virtual body although they were not conscious of the visual stimuli. Finally, it does not seem that one needs to be able to identify an object for multisensory integration to occur. This is well illustrated by the case of a patient with visual agnosia, who had difficulty recognizing everyday life objects by vision but not by touch (Takaiwa et al., 2003). What is interesting is what happened when she used both tactile and visual information simultaneously. If multisensory binding involved conceptual identification of the common source, there should be no interaction between vision and touch in her case, since she could not visually identify the object, and her performance should be the same as when she used only touch to recognize objects. Yet she had greater difficulty recognizing them. In other words, visual information impaired her tactile processing. Hence, one can have a visual experience of an unknown shape, explore it tactually, and bind visual information and tactile information. Multisensory binding can dispense with categorizing the object.

6.4.4 A Nonconceptual Assumption of Unity

The rejection of the sortal hypothesis does not entail that it is not important for the perceptual system to assume that the information from the different sensory modalities comes from a unique source. This is known as the "assumption of unity" in the multimodal literature (Welch and Warren, 1980). The rejection of the sortal hypothesis only entails that the assumption of unity does not require the application of sortal concepts.

In the literature, the assumption of unity is sometimes presented as if it occurs at a high cognitive level:

An intersensory conflict can be registered as such only if the two sensory modalities are providing information about a sensory situation that *the observer has strong reasons to believe* (not necessarily consciously) signifies a single (unitary) distal object or event. (Welch, 1999, p. 373, my emphasis)

It should be noted that even the cognitive interpretation of the unity assumption does not require sortal concepts: the perceptual system only needs to "believe" that there was a single source to the information coming from distinct modalities, it does need to identify it. However, there is no need to adopt a cognitive interpretation of the unity assumption:

A necessary condition for the occurrence of intersensory bias is the *registration, at some level*, that the two discrepant sources of information are emanating from a single event. (Welch and Warren, 1980, p. 649, my emphasis)

We can then suggest an alternative hypothesis of object-based binding, which is less cognitively demanding. Rather than conceptually identifying the source, one can merely perceptually compare objects.

The nonconceptual hypothesis: The perceptual system binds together the sensory states that carry information about the same object. The sameness of the object is determined through its nonconceptual individuation.[11]

Perceptual objects can be grasped in a nonconceptual way (Clark, 2004; Raftopoulos, 2009; Pylyshyn, 2001; 2003, for instance). They are possibly registered in object files: when one perceives an object, one opens a file on it that records its perceptual features. Multisensory binding then depends on the comparison between the visual object file and the tactile object file, for instance. This only involves perceptual similarity between what is perceived by the distinct sensory modalities, instead of its conceptual identity. The perceptual features that are compared include not only location but also size, shape, intensity, motion, orientation, texture, and so forth. They also include temporal properties, such as the time of the onset, of the offset, and the rhythm. The more congruent features there are, the more likely the information from the visual file and from the tactile file concerns the same object. The unity assumption also depends on the prior experience of the coupling of the signals in the world, that is, how often the

---

[11] In Vignemont (2014b), I appealed to the notion of proto-object, but this notion remains obscure and I prefer to use the notion of perceptual object.

two signals are caused by the same source in general. For example, the perceptual system has accumulated information about the statistics of proprioceptive–visual events in the environment: one feels and sees one's hand most of the time at the same location. If the two signals have always been consistent, then the expectation is that they will be highly consistent in the future. Prior experience thus helps to give a causal structure to specific scenes and events (Ernst, 2006; Shams and Kim, 2010).

The nonconceptual hypothesis can then account for the phenomenon of the phantom hand without the cost of positing the use of sortal concepts. It can also account for cases of integrative binding with no spatial link, as in prism adaptation. The visual object file and the proprioceptive object file are assigned to a common source individuated by features rather than by location. Furthermore, the nonconceptual hypothesis removes a potential threat of circularity. In a nutshell, one does not need to identify the object in order to bind the information together so that one can identify the object. Because it does not involve a prior stage of identification, multisensory binding can play a role in the recognition of the object. One then finds out what the object is on the basis of the multimodal experience. For example, in the RHI, participants do not identify the rubber hand as their hand and then feel touch on the rubber hand; instead, they feel touch on the rubber hand and only then do they feel as if the rubber hand were their hand. The identification of the rubber hand as one's own is not a prerequisite for visuo-somatosensory binding; it is a consequence of it.[12] Another asset of the nonconceptual hypothesis is that it removes another potential worry that a more cognitive hypothesis faces. Since Fodor (1983), it has been traditionally accepted that perceptual systems are informationally encapsulated modules, that is, they are insensitive to beliefs. This encapsulation explains why visual illusions can be cognitively impenetrable. One may then argue that if multisensory binding required something like conceptual identification, then multisensory binding could not occur at the perceptual level, but only later at a post-perceptual stage. The only way to avoid this conclusion is to accept a kind of cognitive penetration of perceptual processes

---

[12] Still it is worth noting that under the assumption of ownership, multisensory binding is facilitated: if one is aware that the sensory inputs come from the same source on the basis of other grounds, then one is more likely to integrate the information together even in cases of slight mismatch (Maselli et al., 2016).

(see Macpherson, 2012, for instance).[13] By contrast, a proponent of the nonconceptual hypothesis does not have to settle the complex debate of cognitive penetration. The nonconceptual hypothesis does not involve top-down processes. Hence, the whole process of multisensory binding, from the selection of information to the integration, can occur at the perceptual level.

To conclude, I have argued that the multisensory binding parameter is at the intermediary level between localization and conceptual identification, the level of perceptual object. Integrative binding requires a nonconceptual individuation of the common source of the perceptual experiences. It is thus no threat for bodily IEM. Because it does not involve identification, it does not allow for error through misidentification. Despite the pervasive role of vision, bodily experiences still involve a privileged relationship to one's own body, a relationship that one has with respect to no other bodies.

6.5 Conclusion

According to a classic conception, the content of bodily experiences is exhausted by the information coming from the various bodily senses. In Chapters 4 and 5, I argued against this classic conception, showing the structuring role played by a representation of the long-term properties of the body. In this chapter, I argued that vision is required to maximize the veridical perception of the body. Consequently, bodily experiences in those who have never seen are of a different kind from the way that one normally experiences one's body. Whether or not one is currently seeing one's body, vision plays an essential role in delineating the boundaries of the body, in locating our body parts in space, and in bridging the gap between what happens on the skin and what happens in the external world. In this sense, the bodily experiences of the sighted (or those who were once sighted) may be said to be constitutively multimodal.

As argued, the contribution of vision to the body map is generally advantageous but one may wonder whether it does not have collateral damage. Vision can take many bodies as objects and since vision can

---

[13] A classic experimental finding in favour of this view is the fact that what participants believe about the typical colour of objects affects how they experience the colour (Delk and Fillenbaum, 1965, Olkkonnen et al., 2008).

affect bodily experiences, it can be the vision of one's body as well as the vision of other people's bodies that affect bodily experiences. The function of multisensory binding is to avoid the integration between tactile and proprioceptive information originating from one's own body and visual information originating from other bodies. However, we have seen that such errors can happen. In Chapter 7, we shall further see that the representation of one's own body and the representation of other bodies can have much in common. There are what may be called interpersonal body representations, which indifferently represent bodies, no matter whose bodies they are.

7

My Body Among Other Bodies

Watching someone eating, I would taste and feel their food in my mouth, and I struggled with weight loss because I always felt full.

(Torrance, 2011)

Torrance describes how she can catch, so to speak, other people's experiences: she feels what they feel. She presents an extreme version of what is known as mirror-touch synaesthesia (i.e. tactile sensations experienced on one's own body when one sees another person being touched). Torrance's report may almost seem unbelievable: how could one feel from the inside someone else's sensations? Yet it has been repeatedly shown that in any individual (with or without mirror-touch synaesthesia) the same brain resources can be used to represent one's own and other people's actions, sensations, and emotions (e.g. Keysers et al., 2004; Rizzolatti et al., 1995; Singer et al., 2004). In light of these neuroscientific findings, it has been suggested that there are motor, somatosensory, and affective representations that are "shared" between self and others (Gallese, 2001; Goldman, 2006). One may then wonder whether such interpersonal representations completely erase the distinction between self and others. This question is generally asked while assessing the implications of interpersonal representations for social awareness (e.g. Goldman, 2006; Jacob, 2008). But they may also have implications for self-awareness. Jeannerod, for instance, has argued that since motor representations are shared between self and others, one needs a specific mechanism to determine who is acting, what he calls a "Who" system (Georgieff and Jeannerod, 1998; Jeannerod and Pacherie, 2004):

If, however, we can be aware of both our intentions and those of others in the same way, namely as unattributed or 'naked' intentions, the problem of self-other discrimination does indeed arise.   (Jeannerod and Pacherie, 2004, pp. 139–40)

The argument, which I shall call the naked argument, can be reconstructed as follows:

   (i) There are interpersonal representations;
  (ii) The content of interpersonal representations is impersonal (or "naked");
 (iii) Impersonal contents do not suffice to distinguish self and others;
 (iv) Thus, one can confuse self and others;
  (v) Thus, one needs a specific system to discriminate between self and others.

Jeannerod and Pacherie propose that delusions of control in schizophrenia (i.e. patients feel that their actions are caused by an external force) result from the existence of shared representations of action. Likewise, Ishida et al. (2005) suggest that the fact that somatoparaphrenic patients not only disown their hand but also ascribe it to other individuals is the consequence of the existence of shared representations of the body. One may then ask: do we need a "Whose" system for the sense of bodily ownership in the same way we need a "Who" system for the sense of agency?

7.1 The Body as a Common Currency

The first question to ask is whether we have interpersonal representations of the body as such (Ishida et al., 2015; Vignemont, 2014c). Interpersonal motor, affective, and somatosensory representations have been taken as evidence in favour of embodied social cognition (Gallese, 2007; Gallagher, 2005). But to say that social cognition is embodied is one thing and to say that it exploits body representations is another (Goldman and Vignemont, 2009), and only the latter is of interest for us here. Consider, for instance, the case of imitation. It raises what is known as the intersubjective correspondence problem: how does one map another person's movements to one's own movements (Goldenberg, 1995; Heyes, 2001)? The classic way to approach it has been in *sensorimotor* terms: how does one map visual information to motor commands (Brass and Heyes, 2005)? Alternatively, one can consider that the main challenge is *intermodal* correspondence: how does one map visual information to somatosensory information (Meltzoff and Moore, 1995)? However, the question that interests us here is whether there is also a *bodily* correspondence problem: how does one map representations of another's body onto representations of one's own body? In other words, what bodily parameters, if any, do interpersonal representations encode?

7.1.1 Bodily Mirroring in Action

In 1992, researchers in Parma reported the existence of neurons that fired both when a monkey was grasping a peanut and when it was watching the experimenter grasping it. They named them "mirror neurons" because they reflect other people's actions (di Pellegrino et al., 1992). In humans, action observation and action execution also activate overlapping brain areas (Grezes and Decety, 2001). It was then argued that the same internal representations of action are shared between self and other. But what exactly is shared? Most mirror neurons are only "broadly congruent": they encode the motor goal but not the specific bodily movements (Csibra, 2007). For example, the same movement using *different bodily effectors* (grasping with the mouth and with the hand; grasping food with one's hand and with a stick) activates the *same brain areas* (Gallese et al., 1996; Ferrari et al., 2005). On the other hand, *the same bodily movement* performed with different intentions (grasping a mug for drinking or for cleaning) activates *distinct brain areas* (Iacoboni et al., 2005). What is represented, and thus shared, is a goal rather than a limb. The majority of mirror neurons actually correspond more to emulation (as in taking an umbrella because someone else takes one, for instance) than imitation (as in copying a specific dance step). Hence, action mirroring does not need bodily coding and intersubjective correspondence can be achieved exclusively in motor terms.

However, this is not true in all situations. In particular, imitation of meaningless actions cannot dispense with bodily coding. When one sees a person putting her thumb below her eyebrow, the only solution is to use knowledge about the body to decode the other's movements in terms of the body parts that are seen and in terms of their spatial relations (Goldenberg, 2009). If such bodily coding is impaired, as in ideomotor apraxia, one becomes unable to imitate meaningless gestures (Goldenberg, 1995; Buxbaum et al., 2000; Schwoebel and Coslett, 2005).

Body part coding reduces the visual appearance of the demonstrated gestures to simple spatial relationships between a limited set of discrete body parts. Body part coding facilitates imitation because it produces equivalence between demonstration and imitation that is independent of the different modalities and perspectives of perceiving one's own and other persons' bodies, and because it reduces the load on working memory in which the shape of the gesture must be held until motor execution is completed.   (Goldenberg, 2009, p. 1455)

Further evidence of the involvement of bodily information can be found in our permanent tendency to automatically imitate other people's movements (Brass et al., 2001; Gillmeister et al., 2008). For example, while observing someone moving her index finger, one is faster in moving one's own index finger than in moving one's middle finger. Even the mere observation of coloured patches on another person's static body parts (head, hand, or foot) can suffice to prime action with the same body parts, a finding that is taken as evidence of a "body schema that represents locations on the observer's body and on the bodies of others *in a common format*" (Bach et al., 2007, p. 515, my emphasis). This commonality is confirmed at the neural level. For example, observing hand, foot, or mouth movements performed by another individual selectively activates brain areas for hand, foot, and mouth in distinct regions of the ventral premotor and parietal cortex (Buccino et al., 2001; Wheaton et al., 2004). Correspondence between self and others in action can thus be encoded in bodily terms, in some situations at least.

One of these situations is joint action. It has been found that some effects that normally occur between the two hands of the same individual can occur between the hands of two partners in a joint action such as sawing an object with each using one hand only, thus demonstrating the existence of what has been called a "joint body schema" (Soliman et al., 2015). For example, the attempt to draw a straight line by one partner was affected by the observation of the other partner drawing ovals. Interestingly, joint action also had sensory effects: visual stimuli close to one partner interfered with tactile stimuli on the other partner. Interpersonal body representations might thus go beyond the realm of actions.

7.1.2 Vicarious Bodily Sensations

It suffices to see another person's hand crushed by a door to immediately tense and shrink as if it were our own fingers. But to what extent do we actually feel the vicarious sensation of pain *in our own hand*? A sensation is vicarious if it is caused by another person undergoing a specific sensation and if it is isomorphic to the other person's sensation, but it is so only to a certain extent. The critical question is whether the similarity is also at the bodily level.

To answer this question we need to draw a distinction between emotional contagion and empathy (Vignemont and Jacob, 2012, 2016). Both involve vicarious experiences, but they are of different types. What primarily drives *empathy* is the evaluative component of the emotion, but not the bodily feelings associated with it. In empathetic pain, one experiences a vicarious feeling of the type "it hurts", which is relatively unspecific and spatially indeterminate. It corresponds to what is known as the affective component of the pain matrix (involving the anterior insula, the anterior cingulate cortex, the thalamus, and the brain stem), which encodes the unpleasantness of the painful experience. Empathetic pain is indifferent to the bodily location of pain, and most brain imaging studies on empathy report a selective activation of the affective component only with no activation of the somatotopically organized primary somatosensory cortex SI (e.g. Singer et al., 2004). Hence, the same brain areas can be activated whether the hand or the foot is injured (Jackson et al., 2005).

By contrast, *emotional contagion*, in which one automatically catches another individual's affective state, recruits interpersonal body representations. Interestingly, most instances of emotional contagion are described in bodily terms rather than in affective terms: one talks of contagious crying or contagious laughter instead of contagious distress or contagious happiness. Similarly, experiences of contagious pain are vicarious experiences of the most bodily aspects of pain. They involve what is known as the sensory-discriminative component of the pain matrix (including SI, SII, and posterior insula), which encodes the intensity of pain and its bodily location. Hence, in pain contagion one responds to the perception of another's bodily part subjected to painful stimulation by expecting specific sensorimotor consequences of pain at the same location on one's own body. For instance, seeing the back of another's hand being deeply penetrated by a needle activates the somatotopically organized SI (Bufalari et al., 2007). Motor responses to contagious pain are even muscle-specific: when one sees another's hand being hurt, one's hand muscles automatically freeze, as if it were one's own hand that was injured (Avenanti et al., 2005; 2009).

Although less vivid, one can also have vicarious tactile sensations. In monkeys bimodal neurons that are activated by tactile stimuli on the animal's body can be activated also by visual stimuli on the experimenter's corresponding body parts (Ishida et al., 2010). In humans too, when one sees another person being touched, one activates parts of the tactile areas normally activated when one is touched. However, it is important to distinguish between the active and the passive dimensions of touch, between *touchant* and *touché*. Interestingly, when participants observed another person being touched, several brain imaging studies found activity only in SII, and not in SI (Keysers et al., 2004; Ebisch et al., 2008). This indicates that what is primarily taken into account is not the touched body but rather the act of touching, and indeed such tactile activity can be found even when one watches a roll of paper being touched (Keysers et al., 2004, p. 339): "What is being touched does not matter as long as touch occurs".

Whereas in vicarious active touch one imagines touching something, in vicarious passive touch one imagines feeling touch on one's body. This is especially salient in individuals with mirror-touch synaesthesia, who experience a sensation of touch on their own cheek upon seeing another person being touched on the left cheek for example. If, at the same time, they are touched on the right cheek, they report feeling touch on both sides or they report feeling touch on the left (Banissy and Ward, 2007). Compared to normal subjects they show more intense activity in the somatosensory cortex when they see people being touched (Blakemore et al., 2005). Arguably, individuals with mirror-tactile synaesthesia exploit interpersonal body representations in order to map the location of the tactile stimulation of another's body onto their own body (Ward and Banissy, 2015). The interpersonal dimension of body representations in the sensory domain is confirmed by results in the multisensory literature. Visual enhancement of tactile acuity, as described in Chapter 6, is as effective when seeing one's own body part as when seeing another person's body part (Haggard, 2006). Along the same lines, Thomas et al. (2006) found that participants were faster in detecting touch when they first saw a non-predictive visual cue on another person's body at the corresponding location.

To recapitulate, in some situations, one can map the other onto oneself while bracketing the spatial properties of the body and exploit "disembodied" motoric or affective representations. However, there are other cases in which it can be useful, or even necessary, to analyse other people's movements and sensations in terms of the body parts that move

Table 7.1. Interpersonal body representations

ACTION	Imitation	Embodied
	Emulation	Disembodied
PAIN	Contagious pain	Embodied
	Empathetic pain	Disembodied
TOUCH	Vicarious *touché*	Embodied
	Vicarious *touchant*	Disembodied

and that are stimulated (see Table 7.1). In order to respect bodily congruency, imitation and vicarious bodily sensations can exploit interpersonal body representations (Ishida et al., 2015). The body is then the common "currency" between self and others.

7.2 Two Models of Interpersonal Representations

What exactly is involved in interpersonal representations? I will consider here two main models, which I respectively call the associationist model and the redeployment model. According to the *associationist model*, intersubjective correspondence is achieved by automatic coupling of visual representations and motor/somatosensory representations (Heyes, 2001). By contrast, according to the *redeployment model*, intersubjective correspondence is achieved by recycling the representations originally established for guiding one's actions and framing one's sensations for other people's actions and sensations (Goldman, 2012). As we shall see, each model makes a specific prediction about the content of interpersonal representations, whether it is impersonal or not.

7.2.1 The Associationist Model

Consider first the associationist model, and more specifically the general theoretical framework of the Associative Sequence Learning (ASL) model (Heyes, 2001). According to the ASL model, imitation is based on past experiences of the systematic coupling between the action that one performs and its sensory consequences. Thanks to ASL, one learns sensorimotor associations so that when one sees another person moving, the sensory input can automatically elicit the associated motor output. For example, when I see you raising your finger, this activates a visual representation with the content <a finger raising>, which in turn activates by association the intention <I raise my finger>, which induces the automatic tendency to raise my finger. The association has been built on the basis of past experiences of seeing my finger raising when I had the intention to raise my finger. But it has generalized to any raising finger. On the ASL model, motor representations can represent exclusively one's own intentions. It is only through their association with visual representations, which can represent either one's own body or another person's body, that they are activated by other people's actions. In the multisensory literature, Thomas et al. (2006) also define interpersonal body representations in associationist terms. On their view, there is "a special, automatic mechanism for associating sensory body events" (p. 327). Likewise, Gallagher and Meltzoff (1996, pp. 225–6) suggest, "there is a coupling between self and other, and this coupling does not involve a confused experience". On this model, there is no real sharing, and thus no need for impersonal content. Therefore, there should be no need of a specific mechanism for distinguishing one's body from other people's bodies.

However, one may challenge Gallagher's and Meltzoff's conclusion: coupling can lead to confusion if it is mandatory. If the activation of one representation automatically and systematically induces the activation of the other, and vice versa, then the coupled representations behave as if they constituted a unique representation whose function is no longer self-specific.[1] It would be like marriage: after a while, you can never invite one spouse without the other; they have lost their individuality. Rather than describe this relationship in terms of association, it would instead be better to think of it in terms of fusion, which can lead to confusion. If the function of the associated sensory events is to be activated both by one's body and by other people's bodies, then the activation of the couple can no longer suffice to discriminate among bodies. The critical question is thus whether the coupling is mandatory or not. If it is, then it does not make much functional difference whether there is a single representation or an association of representations.

To settle the debate, one must look at the empirical data but the evidence is mixed. On the one hand, some findings indicate that the associative sequences allow for some flexibility and plasticity and even

---

[1] I would like to thank Wolfgang Prinz for this objection.

automatic imitation can be sensitive to sensorimotor learning (Cook et al., 2010). For instance, body part priming in imitation is reduced following the repeated exposure to incongruent sensorimotor associations, such as observing a hand action while performing a movement with the foot. On the other hand, other findings indicate a tight coupling, so tight that there are bidirectional relations between the two terms of the association. The evidence so far has shown that the perception of other bodies affects the representation of one's own body (other-to-self), but the effect can go the other way around: the representation of one's own body can affect the perception of other bodies (self-to-other). For instance, one is more efficient in detecting changes in another individual's leg posture if one moves one's legs, and more efficient in detecting changes in her arm posture if one moves one's arms (Reed and Farah, 1995). The coupling between self and others can actually be found very early on (Marshall and Meltzoff, 2015). For instance, Longhi et al. (2015) showed that 24- to 48-hr-old newborns looked longer to videos of biomechanically impossible hand gestures than videos of possible ones. They explained this effect of surprise by the sensorimotor expertise acquired in the uterus by the fetuses who were able to make full hand closures and grasping at the third trimester of gestation. One might then be tempted by an alternative model of interpersonal body representations, which appeals to the notion of redeployment.

7.2.2 The Redeployment Model

> Evolution does not produce novelties from scratch. It works on what already exists, either transforming a system to give it new functions, or combining several systems to produce a new one.
>
> (Jacob, 1977, p. 1164)

The interpersonal use of representations that were originally constructed for self-specifying purposes is in line with a general conception of evolution, according to which it co-opts or recycles existing resources for new uses (Gould and Vrba, 1982; Jacob, 1977). There are several versions of it, but here is one way to spell it out, which appeals to the evolutionary notion of exaptation. Both adaptation and exaptation increase the fitness of the organism to its environment, but adaptation results from natural selection, whereas exaptation does not. A now classic example of exaptation concerns the role of feathers in birds, which would have been originally selected due to their role in thermoregulation and only later co-opted for flight.

Some cognitive abilities, and maybe most of them, can then be conceived of as exaptations of existing brain resources (Anderson, 2007, 2010; Dehaene and Cohen, 2007): brain regions are not dedicated to a single task but are recycled to support numerous cognitive abilities. Recycling makes sense from an evolutionary perspective insofar as it is more parsimonious than developing new neural systems. Goldman (2012) applies the notion of redeployment to interpret mirror systems and vicarious sensations. He argues that some brain regions were originally selected to represent the self, and later co-opted for representing other people. On this view, the intersubjective use of body representations increases the current fitness of the organism, but body representations were not originally designed by evolution to accomplish this role.

Although interesting, the redeployment hypothesis leaves many details out. One main concern is the function or functions of the representations that are used and reused. A first possibility needs to be ruled out. Interpersonal body representations have not lost their original function: their use is not confined to social context. On the contrary, the brain areas that are activated in social perception partly overlap with the brain areas activated in non-social context: they are of direct use for representing one's own body, for guiding one's own bodily movements and for providing a spatial frame of reference for one's own bodily sensations.

One may then ask the reverse question: have interpersonal body representations acquired a new function? In their original paper in which they introduced the term "exaptation", Gould and Vrba (1982, p. 6) claim: "adaptations have functions; exaptations have effects". On this view, the intersubjective use of interpersonal body representations does not alter their original biological function, which is to track the state of one's own body only. If so, interpersonal body representations are self-specific. Intersubjective correspondence is then achieved by mapping what happens to other bodies onto the representation of one's own body. This "like-me" view rests on the argument by analogy: it is a good solution to understand other bodies by considering what is true of my body because other bodies are like my own body (Meltzoff, 2007). As in the associationist model, this interpretation of the redeployment model would have no consequence for self-awareness: the interpersonal use does not alter the self-specificity of the content. The problem is that it leads to some unwelcome consequences. In particular, it raises difficulties for a teleosemantic theory of the mind, according to which truth conditions of mental representations are determined by their function. If the sole function of interpersonal body representation were to target one's own body only, then they could be evaluated only relative to it and they could not misrepresent other people's bodies because it would not be their function to represent them. The redeployment model actually does not commit us to accept Gould's notion of biological function. Discussion of the definition of function far exceeds this chapter but we can note that Millikan (1999), one of the main proponents of the teleosemantic approach, proposes a looser notion of proper function. On her view, co-opted processes can have proper functions even if they were not originally designed by evolution. A similar view can be found in Godfrey-Smith's "modern history" approach to functions:

Biological functions are dispositions or effects a trait has which explain the recent maintenance of the trait under natural selection. This is the "modern history" approach to functions. The approach is historical because to ascribe a function is to make a claim about the past, but the relevant past is the recent past, modern history rather than ancient.   (Godfrey-Smith, 1994, p. 344)

On this view, redeployed processes combine their original function with a function that is newly acquired through redeployment (Anderson, 2010; Goldman, 2012). It then seems unlikely that the function of interpersonal body representations is to represent one's body only. Rather, their function is to represent indifferently one's body and other people's bodies: they must have impersonal content.

7.3 The Challenge of Impersonal Content

That we exploit body representations in some social contexts is an interesting fact in its own right. But this also has implications about the content of such representations. We have seen that because they are interpersonal, they are *impersonal*. If we follow the naked argument described in the introduction, one should then conclude that because they are impersonal, they cannot suffice to distinguish one's own body from other bodies. This may be problematic insofar as impersonal body representations may be more pervasive than expected, especially since they can be described both in the perceptual and in the motor domains. They may be so pervasive that one may actually wonder whether there is any body representation whose function is to carry information exclusively about one's own body, and that is thus susceptible to ground the sense of bodily ownership. Hence, although interpersonal body representations are a clear asset for social cognition, one may wonder whether they do not have a cost for the sense of bodily ownership. However, I will argue that their implications for self-awareness are actually limited.

7.3.1 Beyond Impersonal Content

The naked argument works only in the complete absence of bodily information that is self-specific. What the evidence shows, however, is more modest. In brief, it is not because one can experience vicarious pain that pain is impersonal. Our bodily experiences still inform us only about our own body and even if it is true that there are some aspects of the body map that can be shared between self and others, and thus impersonal, those aspects are actually limited. The fact is that impersonal content must be relatively rough-grained, for we can imitate each other despite differences in bodily shape, posture, limb size, muscle strength, and joint flexibility. The brain must thus abstract from major bodily differences. It then seems that what remains in common between all bodies is the configuration of the various body parts, i.e. the fact that we have two hands and two feet and that they are respectively located at the end of our arms and legs. Hence, what is shared is only the representation of the rough structure of the body. This rough structure needs to be filled in for a full-fleshed body map, including information about body metrics, which is highly specific and can hardly be shared.

It is worth noting here that neuroimaging studies have never found a perfect identity between activation for the self and activation for others and one should not overemphasize the extent of sharing. Some low-level processing of bodily information constitutes a common resource between self and other people, which can in turn be used for one or the other. But what is shared—and thus impersonal—is only one component of more complex representations of the body (Ishida et al., 2015). The existence of a self-specific body map is further confirmed by evidence of implicit self-advantage in body recognition (Frassinetti et al., 2008, 2011). Participants are asked to match pictures of body parts. Their performance improves when the pictures display their own body parts compared to when they display other people's body parts from the same perspective.

Hence, they were better in visually processing their own body than any other bodies. Furthermore, it has been found that children with right brain damage were impaired in the recognition of self but not in the recognition of other people's body parts, showing a self-disadvantage, whereas children with left brain damage showed the reverse pattern (i.e. they were impaired in others' but not in self body parts processing) (Frassinetti et al., 2012). Further dissociations between the representation of one's body and the representation of other bodies can be found. In one study, patients with anorexia nervosa were asked to imagine walking through a door-like aperture and then to judge whether they would be able to walk at a normal speed without turning sideways (Guardia et al., 2012). Alternatively, they were asked to imagine another person of the same weight walking through the aperture and to judge whether she could pass. It has been found that the patients mistakenly judged that they could not pass through apertures in which they accurately judged that other people could pass. In other words, their representation of their own body was impaired with no consequence for the representation of other people's bodies. Conversely, patients with heterotopagnosia have selective difficulty locating another individual's body parts on her body, but no difficulty on their own body (Felician et al., 2003; Auclair et al., 2009).[2]

7.3.2 A "Whose" System?

We have seen that even if some components are impersonal, they do not exhaust the content of the body map. The richness of its content goes far beyond what can be shared between self and others. Does that mean that there is no "Whose" system? Let us go back to the comparison with the sense of agency. In the recent philosophical and empirical literature on agency, the prominent view known as the comparator model is that something like a "Who" system involves a mechanism that compares the prediction of the sensory consequences of one's actions and their actual consequences (for a review, see Bayne and Pacherie, 2007). More recently, it has been suggested that the comparator model was too simple and that a Bayesian model was more fruitful to account for top-down influences (e.g. Moore and Fletcher, 2012). Likewise one can find several

---

[2] On the other hand, patients with autotopagnosia fail to locate body parts not only on their own body, but also on other people's bodies, on mannequins, and on drawings of a human body.

attempts of what looks like comparator models of ownership.[3] In the case of agency, the different types of information to compare are efferent and afferent information, but that cannot work in the case of ownership because one can feel one's body as one's own although one is not moving. Instead, it has been suggested that *intermodal* matching plays a key role for ownership (e.g. Makin et al., 2008; Blanke et al., 2015; Tsakiris, 2010; Rochat, 1998).

> It has been proposed that the body is distinguished from other objects as belonging to the self by its participation in specific forms of intermodal perceptual correlation [...] While the rubber hand illusion does not tell us precisely what ingredient might make only certain forms of intermodal correlation relevant to the self, it does show that intermodal matching can be sufficient for self-attribution.   (Botvinick and Cohen, 1998, p. 756)

As for the sense of agency, recent models have provided a more complex story in Bayesian terms, which includes not only multisensory integration but also prior, efferent information, and interoceptive signals (Apps and Tsakiris, 2013; Hohwy and Paton, 2010; Samad et al., 2015).

It is not fully clear, however, what function these integrative mechanisms perform. There are indeed differences between agency and ownership. Each movement is new and thus for each of them one needs to determine whether one is at its origin or not. By contrast, our body is normally always the same. There does not seem to be the need to constantly recompute whose body is mine in everyday life. Even if one can integrate a rubber hand that is merely seen for a couple of minutes, one should not forget that only some subjects experience the illusion. As for long-term bodily modification, the body map can be almost too rigid and remain insensitive to alterations, as illustrated by the phenomenon of phantom limbs, and when amputees receive hand allograft, they report at first being "horrified at seeing the long-expected hand graft" (Zhu et al., 2002): they seldom look at it and even turn their head to the other side while sleeping. Hence, there is some plasticity but it is limited. From the point of view of cognitive parsimony, it thus makes sense that we keep a template of our "habitual body", to borrow Merleau-Ponty's phrase—a

---

[3] For instance, Tsakiris (2010) posits three comparators: between the visual form of the viewed object and a pre-existing body model, between the current state of the body and the postural and anatomical features of the body part that is to be experienced as one's own, and between the vision of touch and the felt touch and their respective reference frames. See also Garbarini et al. (2017) for a comparator model of ownership.

default body map, whose function is to track one's own biological body, and whose content does not change much over time. I thus propose that if "Whose" system there is, it does not play the same role as the "Who" system: one does not constantly compute whose body is mine in the same way as one computes whose action is mine. Does that mean that there is nothing more to explain once we have the default body map in our toolkit? Unfortunately, the reply must be negative. One still needs to explore the precise nature of the default body map and how it is built up. One further needs to investigate what factors can alter it. It is at this point that the Bayesian models that have been described may play a role.

7.4 Conclusion

In the previous chapters I proposed that the sense of ownership is grounded in a frame of reference of bodily experiences that is given by the body map. This role requires the body map to be self-specific: its function must be to represent only one's own body. In this chapter we have examined a challenge to the alleged self-specificity of the body map that is posed by interpersonal body representations. If solving the problem of intersubjective correspondence required the body map, then the body map would not be self-specific and could not ground ownership. However, I have shown that bodily content that is interpersonal, and thus impersonal, does not exhaust the body map. Hence, the function of the body map is not to represent indifferently one's body and other bodies. It is to represent one's own body only and it is thus able to ground the sense of bodily ownership.

The question now is which body map is a good candidate for ownership. Some suggest the body schema, while others favour the body image, but as I will argue in Chapter 8, these notions often remain too vague. What they indicate, however, is that there may be more than one format to encode bodily information. This would not be surprising, given the variety of ways we have of relating to our bodies (e.g. through touch, vision, proprioception, motor behaviour, semantic understanding, affective attitude, etc.). Depending on the bodily aspect, the representation of the body is viewed sometimes as conscious, sometimes as unconscious, sometimes as conceptual, sometimes as nonconceptual, sometimes as dynamic, sometimes as static, sometimes as innate, sometimes as acquired, and so forth. It is now time to clarify the conceptual landscape of body representations. How many are there? And how are they individuated?

8

Taxonomies of Body Representations

One of the main reasons to postulate multiple types of body representation is to account for the multifaceted nature of the disorders of bodily awareness. I have reviewed a few of them in discussing the sensorimotor approach, but their diversity goes far beyond (see Appendix 2). There seems to be no dimension of bodily awareness that cannot be disrupted and one single type of body representation cannot suffice to account for the diversity of syndromes. There need to be more, but how many? Two? Three? Four? And on the basis of what criteria should one distinguish the different types of body representations?

Although there is now a growing consensus that there are at least two distinct types, the body schema and the body image (Dijkerman and de Haan, 2007; Gallagher, 1986; Head and Holmes, 1911; Paillard, 1980), we shall see that there is little agreement about the definitions of these notions. Here I will stress the difficulties that the current body schema/body image taxonomy faces at both the conceptual and the empirical levels. Some theorists have concluded that the task of classifying body representations is a "slippery issue" to avoid at all cost (Holmes and Spence, 2005, p.16), but what may be problematic is not so much the project of classifying body representations as opposed to attempting to ground conceptual distinctions exclusively on empirical dissociations. After offering a critical analysis of the notions of body schema and body image as they are used in the philosophical and psychological literature, I will propose distinguishing different types of body representations on the basis of their direction of fit and of their spatial organization.

8.1 How Many Body Representations?

In 1911, Head and Holmes present the first taxonomy of the representations of the body. In addition to what they call a *body image*, which constitutes the conscious percept of the body, they distinguish two types of unconscious representations, called *body schemata*: one records every new posture or movement whereas the other maps the surface of the body on which sensations are localized. However, Head's and Holmes's rich taxonomy was soon forgotten. For a long time, the terms of body schema and body image had been used interchangeably (see Schilder, 1935, for instance), whereas the fine-grained distinction between the two types of body schema had almost completely disappeared. The common assumption was that there was a single type of body representation (Schilder, 1935; Berlucchi and Aglioti, 1997; Melzack, 1990; Reed and Farah, 1995), and bodily disorders were called "disturbances of the body schema" in the neurological literature and "disruptions of the body image" in the psychiatric literature (Denes, 1990). However, as noted by Poeck and Orgass (1971, p. 255), "the only obvious common denominator was that they had something to do with the human body".

Nonetheless, since the 1980s there has been a renewal of interest in distinguishing various types of body representation (see Longo, 2017 for review). More specifically, the distinction between body schema and body image has become the stock in trade of much recent work in neuropsychology, cognitive neuroscience, and philosophy (e.g. Gallagher, 1986, 2005; Dijkerman and de Haan, 2007; Paillard, 1999; Berlucchi and Aglioti, 2010; de Preester and Knockaert, 2005; Vignemont, 2010; di Vita et al., 2016). In brief, the body schema is involved in action, whereas the body image expresses how we perceive our body. However, as I shall show, this basic description leaves many questions open.

One may first question the precise definitions of these two types of body representations. The notion of body image has attracted most attention (and controversy) because of its lack of unifying positive definition. The dynamics of the body image varies from short-term to long-term and it is both nonconceptual (body percept and body affect) and conceptual (body concept). How could all these aspects be part of one single category? Because of this apparent heterogeneity, there is a risk that the concept of body image becomes empty of meaning and of explanatory power. The body image seems to be just whatever is left over after we are done talking about the body schema but the problem is that even the notion of body schema is also somewhat obscure: it is generally assumed to be linked to action but little is said about their exact relationship. It seems that body representations activate different brain regions, whether they are involved in action or not (di Vita et al., 2016), but can one offer a more refined and precise description of each notion?

One may further question the fact that there would be only two types of body representations. In line with Head's and Holmes's original taxonomy, a three-fold distinction has been suggested: (i) a sensorimotor body representation for action (also called body schema); (ii) an implicit body model (also called visuo-spatial body representation or body structural description), which can be defined as a stored "reference description of the visual, anatomical and structural properties of the body" (Tsakiris, 2010, p. 707); and (iii) an explicit body image (also called body semantics), which includes the beliefs that one has about one's body (Schwoebel and Coslett 2005; Sirigu et al. 1991; Longo and Haggard, 2010, 2012). However, such taxonomy leaves out what Gallagher (2005) calls the body affect, and more generally what some people call the interoceptive body (Seth, 2013; Tsakiris, 2017). Should one then posit a further type of representation? But when to stop? The challenge here is to find a good balance between an infinite multiplication of body representations and a lack of homogeneity within each type. This requires determining clear criteria on which basis one can individuate the different types. Here is a sample of the characteristics that have been used so far:

- Availability to consciousness (unconscious versus conscious)
- Dynamics (short-term versus long-term)
- Functional role (action versus perception)
- Format (sensorimotor versus visuo-spatial versus conceptual)
- Modality of input (proprioception versus vision versus interoception)
- Bodily information (metrics versus posture versus homeostasis)
- Perspective (first-person versus third-person).

The problem is that what is a distinctive feature for one author can be irrelevant for another. For instance, Head and Holmes (1911) favour availability (body schemata are unconscious unlike body image) and dynamics (postural schema is short-term whereas the schema encoding the surface of the body is long-term); Paillard (1999) exclusively highlights functional considerations (the body schema is for action whereas the body image is for perception); Gallagher (1986) combines availability and functional role (the body schema is an unconscious sensorimotor function whereas the body image is a conscious body percept, affect, and concept). It is no surprise therefore that the very same notion can be ascribed opposite properties by different authors. This also explains how the same pool of evidence can lead to opposite interpretations, as we shall now see.

8.2 Deficits in Body Schema and Body Image

The problem is that the few tests that have been developed to evaluate the different types of body representations are not conclusive and receive contradictory interpretations (see Table 8.1). Therefore, the same neuropsychological disorders can be diagnosed either as a deficit of body schema or as a deficit of body image (see Table 8.2). I shall not review each syndrome here, but will instead focus on the dissociation between personal neglect and deafferentation, which Gallagher (2005) uses to support his distinction between the body schema and the body image. The detailed analysis of these two syndromes will allow me to further highlight the obscurity of the two notions.

Table 8.1. Evaluating the distinct types of body representation

Body schema	Body image	
Sensorimotor	Visuo-spatial	Conceptual
Pointing to one's body parts (Paillard, 1999)	Naming one's body parts Pointing to pictured body parts (Paillard, 1999; Anema et al., 2009)	
Motor imagery (Schwoebel and Coslett, 2005)	Pointing to one's body parts: contiguous errors (Sirigu et al., 1991; Semenza and Goodglass, 1985)	Pointing to one's body parts: functional errors (Sirigu et al., 1991; Semenza and Goodglass, 1985)
	Pointing to one's body parts and to a pictured body (Schwoebel and Coslett, 2005)	Matching body parts to functions and body-related objects (Schwoebel and Coslett, 2005)

Table 8.2. Neuropsychological dissociations and their interpretations

Body schema	Body image deficit	
deficit	Visuo-spatial	Semantic
Deafferentation *(Paillard, 1999; Gallagher, 2005)*	Autotopagnosia *(Sirigu et al., 1991; Schwoebel and Coslett, 2005)*	Autotopagnosia *(Sirigu et al., 1991)*
Personal neglect *(Coslett, 1998)*		
Apraxia *(Buxbaum et al., 2000; Schwoebel and Coslett, 2005)*	Numbsense *(Paillard, 1999)*	Body-specific aphasia *(Schwoebel and Coslett, 2005)*
	Personal neglect *(Gallagher, 2005)*	
	Apraxia *(Goldenberg, 1995)*	

8.2.1 Pointing to What?

Let us first consider how one diagnoses deficits of body schema and body image. One task that has come back again and again since Head and Holmes (1911) is that of asking participants to point to a body part that has been either touched, named, or visually shown on a picture. Although this task has been widely used, especially in neuropsychology, it is not clear what it is supposed to assess. On the one hand, performance in pointing to one's body is taken as a measure of the body schema, which is impaired in deafferentation, but not in numbsense (Gallagher and Cole, 1995; Paillard, 1999; Rossetti et al., 1995; Dijkerman and de Haan, 2007). On the other hand, it is taken as a measure of the body image, which is impaired in autotopagnosia, but not in apraxia (Sirigu et al., 1991; Schwoebel and Coslett, 2005). In addition, sometimes it is a measure of both the visuo-spatial and the conceptual components (Sirigu et al., 1991), sometimes it is a measure of the visuo-spatial body image only (Schwoebel and Coslett,

2005).[1] So, what do pointing errors mean? A disruption of the body schema and/or a disruption of the body image(s)?

The answer is that it can reflect either type of disorder depending on the kind of hand movement being performed. Most of the time, hand movements are instrumental actions, performed because one wants to do something with a specific object, such as pick it up. However, in the experimental context, manual responses can also be used to give perceptual reports (e.g. scaling the distance between the fingers to communicate a judgement about the width of the object). This duality of use is especially salient in the case of pointing responses, for pointing is an unusual action. It is a communicative act used to attract an observer's attention to a target and to indicate it. Pointing to body parts (rather than pointing to objects) is even more special, for it involves representing the body both as the goal and as the effector of the movement and it develops later than most body-directed actions (Poeck and Orgass, 1964). In addition, pointing might recruit different types of body representations depending on the target (e.g. one's hand versus a hand picture), on the type of movements performed (e.g. slow visually guided informative gesture versus fast ballistic movement), and on the type of errors measured (e.g. wrong tactile locus within the right body part versus wrong body part). The weight can then be more on the body schema or on the body image. These differences can explain why pointing can be interpreted in some cases as a measure of body schema (ballistic spatially fine-grained pointing on one's body), and in others as a measure of body image (visually guided categorical pointing on a body picture). Consequently, although pointing is the main experimental tool to investigate body representations, it is neither an exclusive measure (i.e. specific to one kind of body representation only), nor an exhaustive one (i.e. representative of the whole body representation, and not only of some of its aspects).

The relative lack of clinical tools that are specific enough can also be partly explained by the fact that the various types of body representations

---

[1] It has also been argued that it is not a measure of body representation at all but rather a measure of non-verbal communication (Cleret de Langavant et al., 2009). This latter view, however, is supported by evidence on pointing to another individual's body parts. Since the situation is already complex enough without adding intersubjectivity, I focus here on pointing to one's own body parts.

interact all the time, and it is thus empirically difficult to disentangle them. For the same reason, we shall see that it is difficult to find "pure" selective deficits.

8.2.2   Personal Neglect: a Deficit of Body Image?

> (What did you do with the [left] leg?) "With what leg?" [. . .]
> (Where is your left eye?) "Probably I have it in my head." (Show it).
> "I am not certain." He pointed with both hands to his right eye and to the base of his nose.
>
> (Schilder, 1935, pp. 314–15)

Personal neglect is clinically defined by a lack of exploration of half of the body opposite to the right damaged hemisphere (McIntosh et al., 2000). When patients are touched on the left side, they do not report any tactile sensations. The left side of their body does not appear to exist for them. However, they do not complain about the missing half. They feel as they have always felt, as if their body were complete. They merely behave as if the left side did not exist. Personal neglect is often, but not always, associated with extrapersonal neglect, which affects half of the external space (Bisiach et al. 1986; Guariglia and Antonucci, 1992).

One reason for thinking like Gallagher (2005) that personal neglect involves body image deficits is that patients have difficulty in recognizing touched fingers and in reconstructing a body and a face by using pre-cut puzzle pieces (Guariglia and Antonucci, 1992). Such a finding indicates impaired body perception at the explicit level, but physiological measures showed that tactile stimuli could still be *implicitly* processed in a patient with personal neglect who was unable to report tactile stimuli on his left side (Vallar et al., 1991).[2] Hence, if the body image consists in body perception, whether conscious or unconscious, then it is partially preserved, but if it consists only in *conscious* body perception, then it is missing.

On the other hand, the body schema seems to be at least partially impaired (Coslett, 1998; Baas et al., 2011). For example, patients tend to "forget" the left side of their body in reflective actions like combing or shaving, and they often underuse their left limbs. In addition, they are

---

[2] Likewise in *extrapersonal* neglect, explicit spatial processing can be impaired while implicit spatial processing is preserved (Marshall and Halligan, 1988): when patients with extrapersonal neglect were presented a picture of a house with the left side on fire, they were not aware of the fire, but they claimed that they would not live in the house.

impaired in motor imagery, which constitutes an interesting task to assess the integrity of the body schema because it shares many properties with physical actions at the physiological level (muscle activity), at the kinematic level (similar physical constraints and laws), and at the neural level (shared patterns of brain activation) (Jeannerod, 1997; Schwoebel et al., 2002a; Schwoebel and Coslett, 2005).

To summarize, far from being a paradigmatic example of a clear body image deficit, personal neglect probably combines partial impairments of both body image and body schema, possibly resulting from more general spatial and attentional deficits (Bisiach and Berti, 1995; Kinsbourne, 1995).

8.2.3 Peripheral Deafferentation: the "Missing Body Schema"?

As already discussed in the previous chapters, patients after peripheral deafferentation lose all sense of proprioception and touch while their motor nerves are spared. Since they do not know where their limbs are if they close their eyes, they need to learn to exploit visual information in order to calibrate and guide their movements. They are also impaired in localizing thermal stimuli, which they can still perceive, on their own body without vision, but they can localize them on a pictorial representation of their body (Paillard, 1999). Gallagher and Cole (1995) conclude that major parts of the body schema in deafferented patients are "missing" and compensated by a reflexive use of their body image.

But in what sense is the body schema defective in these patients? After all, they can move in a relatively impressive manner. Spending several days with Ginette Lizotte, one of the most famous—and most impaired—patients, as her interpreter, I often forgot that there was anything abnormal besides her wheelchair. She could cut her meat while having a normal discussion at lunch and even gesture with her knife and fork like everybody else, or so it seemed. As noted by Travieso et al. (2007, p. 223, my emphasis):

What we think the results show is that G.L. was simply poor at pointing *without vision*: she was equally good at pointing with vision towards her hand or towards a picture. Her problem was *not with action itself*, but with the use of haptic information alone (without vision).

If action is not the problem, then one may conclude that the body schema is at least partially preserved in deafferentation. To still assume that the body schema is missing requires using a definition of the body schema that does not give a central role to action. For instance, one may claim that the body schema is primarily defined by its sensory input. If the body schema is fed only by proprioception and touch, then it is necessarily impaired in peripheral deafferentation. However, I argued in Chapter 6 in favour of a constant interplay between bodily senses and vision. Not only are bodily experiences multimodal; the body schema is too (Vignemont, 2016; Wong, 2014, 2015). Evidence of this multimodality can be found in the following study: if the visual size of the bodily effector is temporarily distorted, the planning of the subsequent movement is affected (Marino et al., 2010; Bernardi et al., 2013). The role of vision for the body schema is thus not unusual. It is merely more drastically important in the case of deafferented patients.[3] One may then conclude that the body schema is preserved, although based on different weighting of information. Whereas proprioception normally plays an important role, it has been taken over by vision and thermal perception in deafferented patients.

Still one might maintain that the body schema is "missing" because of the specific way actions are achieved: whereas most actions are performed without attention in healthy individuals, they would require reflexive monitoring in deafferented patients (Gallagher and Cole, 1995). This assumption raises an empirical question: do deafferented patients still need to consciously control their movements? It also raises a theoretical question: is it only the body image that is available to consciousness?

Let us first consider what the patients themselves report. It is true that when the disease first struck, the patients had to learn how to visually guide their actions, which required a lot of effort and attention. For example, it took three months for Ian Waterman to visually control his movements. The question is whether more than thirty years later bodily control still requires the same effort. Ginette Lizotte told me that she now felt like an automatic driver: when you learn to drive a car, you have to pay attention all the time to all the details, but after a while, you drive without consciously monitoring the visual information that you receive

---

[3] It is interesting to note that the body schema is less malleable in deafferentation than normally. One deafferented patient could be trained to use a tool to perform a specific movement, but the effect of tool use was not generalized to other movements, thus showing that the tool was not incorporated at the level of her body schema (Cardinali et al., 2016). Without proprioception, it is more difficult to update the body schema.

about the other cars, the road, etc. You still need to process a large amount of visual information, but with practice, your actions become automatic and do not prevent you from simultaneously engaging in other activities. There seems to be no principled reason why the extensive use of visual information made by deafferented patients could not be of this kind.[4]

But even if we granted that deafferented patients needed to consciously monitor their movements, would that suffice to show that they rely on their body image instead of their body schema? It does only if one assumes that by definition only the body image can be conscious (Head and Holmes, 1911; Gallagher, 2005). There is, however, a milder position that is defended by Schwoebel and Coslett (2005). On their view, the body schema is rarely conscious, but it can be so in some circumstances, as in the use of conscious motor imagery.[5] For instance, one task used to assess the integrity of the body schema in patients is to ask them to consciously imagine performing actions: the patients then have conscious access to mental representations of their body in action, and thus, one might say, to their body schema. One may then suggest that deafferented patients simply consciously exploit their body schema, as in conscious motor imagery. On this view, the availability to consciousness is not a criterion that differentiates the body schema from the body image.

Hence, far from being a paradigmatic example of deficit in body schema, deafferentation rather reveals its ability to adjust to the sensory inputs that are available. True, the body schema is altered, relying more on vision than before, but for all that, it is not "missing".

8.2.4 A Constant Interaction

This brief overview of the alleged dissociation between personal neglect and deafferentation illustrates the fact that most bodily disorders do not lead to straightforward diagnosis in terms of either body image deficit or body schema deficit. This can be partly explained by the fact that if body

---

[4] The case of Ian Waterman may be slightly different from the case of Ginette Lizotte because unlike her he was not in a wheelchair and he could walk, which was very demanding for him. However, even very effortful actions can be automatized (complex dance steps, for instance), and it is hard to see why this would have been impossible in his case.

[5] That does not mean that the computations involved in the construction of this representation are available to consciousness.

schema and body image there are, they continuously interact, and thus they are hard to dissociate. As a consequence, deficits of body schema and body image are often intermingled and clear cases of specific disruptions rarely found. This also explains why in our everyday life the body we perceive rarely conflicts with the body we act with: under normal conditions, when perception is not distorted, the body schema and the body image are congruent. It is then no surprise that very few studies have succeeded in explicitly dissociating the two body representations in healthy individuals (e.g. Vignemont et al., 2009; Kammers et al., 2006; Kammers et al., 2009a; Tsakiris et al., 2006). Their interaction can be easily explained. The system tends to decrease errors whenever it is possible. To do so, the two body representations can take advantage of each other: the body image refines its content thanks to information held in the body schema and vice versa. This does not imply that they match perfectly but only that their respective content is tempered in light of the other in order to maximize efficiency and to get them as accurate as possible. The interaction between them can go both ways, as illustrated by the following two studies in healthy individuals.

In Chapter 2, I reported that tool use can modify the estimated length of the arm: one judges one's arm as longer after using a grabber than before (Cardinali et al., 2009a). Insofar as the tool is taken into account by the motor system, it is integrated into the body schema, but what about the body image too? To answer this question, Cardinali and her colleagues (2011) did the following study. After using a long grabber, participants were asked to localize specific targets on their arm (finger tip, wrist, and elbow), which enables computing the estimated size of the arm. There were two types of inputs and two types of outputs (see Table 8.3). The experimenter indicated the bodily target either by touching it (tactile input) or by naming it (verbal input). The participants' response could then be of two types also, either verbal (by indicating the corresponding number on a meter in front of them) or motor (by pointing to the target). The comparison between the two opposite conditions (tactile-motor condition versus purely verbal condition) leads to a clear dissociation: tool use induces arm lengthening in the tactile-motor condition, but not in the purely verbal condition. However, what are interesting are the other "intermediate" conditions: participants showed an effect of arm lengthening even when they replied verbally after tactile stimulation, but they showed no such effect when they

Table 8.3. Effect of tool use on the estimated size of one's arm (Cardinali et al., 2011)

	Tactile input	Verbal input
Verbal response	Effect of tool use	No effect of tool use
Pointing response	Effect of tool use	No effect of tool use

pointed to the target that is named (see Table 8.3). This result again reveals the obscurity of the pointing task. It also reveals that the body schema can reshape the body image.

Other findings using the RHI paradigm illustrate the constant inter-action between the body schema and the body image. One may believe that the rubber hand is integrated only at the level of the body image since participants report no agentive feelings towards the rubber hand (Longo et al., 2008) and since action can be immune to the illusion (Kammers et al., 2009a). However, several studies have also found that motor responses can be sensitive to the RHI in specific contexts (Riemer et al., 2013; Heed et al., 2011, Kalckert and Ehrsson 2012; Holmes et al., 2006; Tieri et al., 2015b). Let us just focus on the following study by Kammers and her colleagues (2010). Unlike the classic version of the RHI, the participant's hand and the rubber hand were shaped as if they were ready to grasp an object and both the index fingers and the thumbs of the real hand and of the rubber hand were stroked. Participants were then asked to grasp an object in front of them. It was found that the kinematics of their movements differed after synchronous and asyn-chronous stroking. More specifically, when they felt the rubber hand as being their own, their own grip aperture was significantly influenced by the size of the seen grip aperture of the rubber hand. This reveals that this time the body image has reshaped the body schema.

8.3 A Practical Photograph of the Body

Spicker (1975, p.182) notes: "We allow ourselves to speak of the body image and other such scheme or ghosts, which, I think, we would well be rid of by adopting a method of intellectual exorcism". We have indeed seen the difficulties that the body schema/body image distinction faces, which can be explained by their constant interaction but not only. There is a lack of precise understanding of the functional role of the body schema as opposed to the body image and without clear definitional criteria, they cannot play any explanatory role. At this stage, we should simply decide that we are better off without them. Instead of providing a new definition for them, which undoubtedly would lead to new disagreements, I thus prefer to discard them and start all over again in order to reconsider the relationship between body representations, action, and bodily experiences. Because it seems to be the most tractable, I shall focus first on the specific type of body representation that is used for planning, guidance, and control of action.

8.3.1 Knowledge of Bodily Affordances

One might expect that a better understanding of action-orientated body representations would emerge from theories of agency. However, most computational models and philosophical theories of action do not even mention the body, although bodily information is required at every step by the motor system. Even the formation of intention involves some knowledge of one's bodily capacities. In addition, when it comes to planning specific movements the motor system needs to take into account not only what the body can do but also the current bodily state. Finally, action control involves comparing the location at which one's body should be with its actual location. The crucial question is: how is bodily information represented by the motor system?

O'Shaughnessy (1980, vol. 1), who is one of the rare people to have acknowledged the crucial role of the body for action (see also Wong, 2015), suggests that we have what he calls a practical photograph of the body:

> In an analogous practical sense we all of us have knowledge of our limb spatial possibilities; so that a man will introduce his hand into a cupboard but will not attempt to insert it into a thimble! Indeed through assembling the lowest common denominator of all the acts he will undertake with his hand, we might finally manage to assemble a sort of "practical photograph" of the hand.
>
> (O'Shaughnessy, 1980, vol. 1, p. 225)

Arguably, the practical photograph of the body has been constructed partly on the basis of past sensorimotor feedback. I have emphasized so far the role of vision, but action also calibrates how one represents one's body. Early on, young infants engage in repetitive actions on their own body and explore visual-proprioceptive correspondence, as they observe their legs kicking (Rochat, 1998; Bremner, 2017). Later in life, sensory feedback that one receives when acting is taken into account by body representations, especially during childhood and adolescence when the body is still growing in size and body representations need to adjust: hitting the shelf with one's head, for instance, indicates that one has become taller than the height of the shelf. Even later, each time that one actively uses a tool, one temporarily incorporates it into one's body representation (Cardinali et al., 2009a).

The "practical photograph of the body" is thus partly grounded in motor expertise, but what are its content and format? One interpretation is in terms of the notion of *affordances*, which was originally developed by Gibson (1979) to describe the specific relationship between an agent and specific properties of the environment. For example, one may say that when one sees a chair, one sees that the chair affords a seat. But sitting on the chair involves not only seeing it as affording a seat but also knowing the bodily movements to perform to sit on it. One may then talk of bodily affordances (Alsmith, 2012; Smith, 2009; Wong, 2009). Knowledge of bodily affordances represents body parts in terms of their capacities for movements. It concerns two types of bodily information, postural and structural. Knowledge about *postural affordances* consist in:

knowing which specific movements of those general kinds are possibilities for one here and now. If one of one's legs is, say, not bent at the knee, straightening it is not one of one's present options.   (McDowell, 2011, p. 142)

Knowledge of postural affordances has a very short life span. It is built up at time t, stored in working memory, and erased at time t + 1 by the next movement. Since action occurs also on a very brief time scale, one may believe that only postural affordances are of direct relevance for guiding action. However, to move one's arm, one needs to know not only its position at time t, but also its size, which has not changed for the last ten years. For example, to switch on the light, you need to know the length of your arm in order to plan how far you should stretch it. Hence, one needs knowledge about long-term structural properties of the body, including bodily configuration, bodily size, flexibility of the joints, and muscle strength, which constitute *structural affordances*:

a familiarity with the possibilities for bodily acting that come with having the kind of body she has: for instance, a familiarity with the different movements that are feasible at different joints [...] for instance that the kind of joint a knee is allows a leg to be bent so as to take the foot to the rear but not to the front or the side.   (McDowell, 2011, p. 142)

It is thanks to the representation of structural affordances that one does not attempt to move in biologically impossible or painful ways. It is also thanks to it that one does not over- or under-reach when trying to get an object.

Although relatively intuitive, one may still wonder to what extent the obscure notion of affordances actually helps in our attempt at clarifying the equally obscure notion of a practical photograph of the body.[6] The notion of affordances carries with it some specific relationship between perception and action, but the nature of this relationship is highly controversial. In particular, one may wonder in what sense representations of bodily affordances depart from purely perceptual representations of the body.

8.3.2 Bodily Pushmi-Pullyu Representations

Let us consider Anscombe's (1957) famous example of the shopping list. The shopping list can be conceived in two different ways. At first, it tells you what to buy. Your shopping is more or less successful depending on whether you have bought all the items on the list. Later, it describes what is in your grocery bag. The description can be more or less accurate depending on whether the list corresponds to what is in your bag. Likewise, mental representations can be either directive with success conditions, or descriptive with truth conditions (Searle, 1983). In the former case, they have world-to-mind direction of fit. For instance, the world must be made to match your desire, and this can be more or less successful. In the latter case, they have mind-to-world direction of fit. For instance, a belief must match, or fit with, the world. The problem with this dichotomy is that it leads to a disconnection between the way the world is and how to act on it. To avoid this, Millikan (1995) argues that in some circumstances, we use what she calls *pushmi-pullyu representations* (hereafter PPRs). Like Dr Doolittle's mythical animal, PPRs face in two directions at once: they have both a mind-to-world and a world-to-mind direction of fit, both truth conditions and success conditions. The

---

[6] For one thing, affordances are classically defined as relational properties between a surface or a shape and the agent's bodily capacities. Should one then understand bodily affordances as the relation between the body and itself? This does not seem to capture the idea. More likely, the hypothesis is that they consist in the bodily side of environmental afforsances.

content of PPRs varies as a direct function of a certain variation in some aspect of the environment that it represents and directly guides behaviour directed towards this aspect. PPRs need no inferential structure for them to have a relation to action: there is no need to translate descriptive information into directive information; it is constitutive of the content, which builds the command for certain behaviours into the representations. As such, PPRs afford great economy in terms of response time and cognitive efforts. A clear example of PPR is a performative act. When the priest declares, "I pronounce you husband and wife", he or she both describes the fact that you are married and he or she makes it happen. Another example provided by Millikan is the perception of affordances: they both describe the properties of the environment and guide your actions towards it.[7] One may then suggest applying the notion of PPRs also to representations of *bodily* affordances.

Although they describe what you *can* do, bodily affordances also tell you what you *should* do, and this is so at several levels. In the now classic computational framework of action, the motor system uses two types of internal models (Figure 8.1): the inverse model and the forward model (Wolpert et al., 2001). The *inverse model* has the role of computing the motor command needed to achieve the desired state given the agent's bodily affordances. The inverse model is thus fed by information about structural and postural affordances. One may even say that the content of the representation of bodily affordances is coercive: it heavily constrains action planning so that one normally cannot help but use it to guide one's bodily movements. In short, knowledge of bodily affordances makes you act *in a certain way*. As summarized by Millikan (1995, p. 191):

The representation of a possibility for action is a directive representation. This is because it actually serves a proper function only if and when it is acted upon. There is no reason to represent what can be done unless this sometimes effects its being done.

In parallel with the inverse model is run the *forward model*, which predicts what the action will be like given the specific body that executes the motor command and allows anticipatory control of movements. Here also, I argue, the representation of postural affordances has a

---

[7] Bayne (2010) and Pacherie (2011) suggest that agentive experiences are also instances of PPRs.

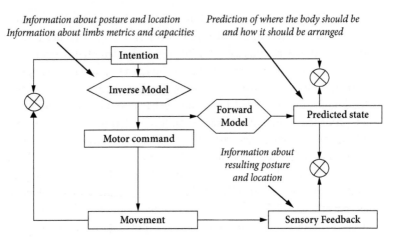

Figure 8.1. Computational model of embodied action.

clear directive function. It represents the desired posture and location. For it to be satisfied, the body must meet this posture and this location. Hence, the direction of fit is world-to-mind: the body must match its representation, and if it does not match it, then the agent must update the motor command.

To conclude, the function of the representations of bodily affordances is to describe the body and to guide action, and it is fulfilled if and only if the motor system obeys them, so to speak. We can now give the following general definition of the body representations that guide action. They are bodily PPRs that describe how the body is in terms of bodily affordances and that direct bodily movements. There is more than one type of bodily PPR, depending on what role the PPRs play for action, whether they are for planning or for online control, whether they represent long-term structural affordances or short-term postural affordances. Action then results from the interplay of these various bodily PPRs.[8]

---

[8] For example, we can imagine a situation in which a child over-reaches a jam pot due to discrepancy between his arm size as represented in his bodily PPR and his actual arm size. He can then adjust his long-term bodily PPR and the next time he will no longer over-reach. Consequently, bodily PPRs involved in control can have effects on bodily PPR involved in planning.

8.4 Is the Body Map Hot?

We now have a better grasp of the type of body representations exploited for planning and guiding action. We know the type of information encoded, namely, postural and structural affordances. We know how this information is encoded, that is, in both descriptive and directive content. We can thus decide whether there need to be other types of body representations and in what ways they differ. In particular, I will consider the type of body representations that grounds the felt localization of bodily experiences. In Chapter 4, I argued against the sensorimotor approach to bodily awareness that reduces the spatiality of bodily experiences to bodily activities. Instead, I claimed, one needs to exploit the representation of the long-term properties of the body, which I call body map, to provide a spatial frame of reference for the felt localization of bodily sensations. One may then defend a weaker version of the sensorimotor approach, which grants the necessity of a body map, but argues that the body map itself is sensorimotor (Bermúdez, 2005, 2017; Brewer, 1995; Smith, 2009; and to some extent, O'Shaughnessy, 1980, vol. 1 and McDowell, 2011). I shall call this view the *practical hypothesis*. However, I will show that the PPR of structural affordances cannot suffice to account for the felt localization of bodily experiences.

8.4.1 The Practical Hypothesis

Consider first kinaesthetic experiences. There is an intuitive sense in which the frame of reference of kinaesthetic experiences has a motoric format: in kinaesthetic experiences one normally experiences one's limbs in terms of the movements that they afford (Wong, 2009). This seems to be confirmed by the fact that kinaesthetic experiences become inaccurate if the representation of structural affordances is inaccurate. Consider the case of anosognosia for hemiplegia, which consists in the lack of awareness of paralysis. Interestingly, anosognosic patients can report that they have raised their paralysed arm, although they have not moved an inch. Their kinaesthetic experiences are thus erroneous and one way to explain their errors is that the representation of their structural affordances has not been updated (Schwoebel et al., 2002a). Kinaesthetic experiences then simply inherit the inaccuracy of the representation of structural affordances.

The awareness of the body in action thus exploits bodily PPRs but one may reply that this is hardly surprising. The crucial question is whether the felt localization of other types of bodily experiences, or even of all of them, exploits them too. Let us now consider the case of painful experiences, for which the link to action is the most salient. Pain has indeed an intrinsic motivational force that immediately causes us to act to prevent or stop the sensation. Furthermore, it has been shown that action-orientated body representations are modified in chronic pain conditions (Schwoebel et al., 2002b; Moseley, 2004). This alteration is necessary to avoid moving the painful body part, for instance (e.g. one should not turn one's head with a stiff neck). Focusing on this motivational dimension, Klein (2015a) argues that pains are bodily commands that are felt to be located within the frame of reference given by what he calls "body schema", and what I call bodily "PPRs":

> To begin, the relevant sense of bodily location seems to depend on something like an internal representation of the body [. . .] pain should have a relationship to the body schema rather than the body image.   (Klein, 2015a, 92–3)

On Klein's view, when I hurt my thumb badly with a hammer, I feel the pain within the frame provided by bodily PPRs and such a localization allows a direct link to the protective movement of putting my thumb in my mouth: there is no need to translate where I feel pain to a motor format for motor control since the localization of the pain sensation is already in a motor format.

More generally, Brewer (1995) argues that all bodily experiences are located in terms of their implications for action:

> The intrinsic spatial content of normal bodily awareness is given directly in terms of practical knowledge how to act in connection with the bodily locations involved. The connection with basic action is absolutely not an extrinsic add-on, only to be recovered from a detached map of the vessel the subject of awareness happens to inhabit, on the basis of experiment and exploration. It is, rather, quite essential to the characterization of the spatiality of bodily sensation.
> (Brewer, 1995, p. 302)

On this view, when I feel the contact of the table on my arm, I feel it on the body part that is represented as being able to do such and such movements and if I were to react, and retrieve my arm for instance, the spatial content of my tactile experience could be a direct input to motor control thanks to its motor format. The practical hypothesis has the advantage of being relatively parsimonious. On this view indeed, there is a unique type of representation of the long-term properties of the body used for both action and the felt location of bodily experiences. There may be a functional distinction at the short-term level, for instance between postural representation for perception and postural representation for action, but the functional distinction does not apply when it comes to representing the enduring properties of the body (i.e. bodily configuration and bodily metrics):

This first category of body-relative information performs two tasks. First, it is responsible for the felt location of sensations. Sensations are referred to specific body-parts in virtue of a body of information about the structure of the body. Second, *the same body of information* informs the motor system about the body-parts that are available to be employed in action

(Bermúdez, 2005, p. 305, my emphasis)

The problem, however, is that the practical hypothesis can hardly account for the empirical dissociations that I presented in Chapter 4. Let me briefly recall them:

(i) The patient RB with allochiria acts on the side of her body that is stimulated, while she systematically reports feeling sensations, including pain, on the other side (Venneri et al., 2012).[9]

(ii) The patient KE fails to accurately localize his sensations when asked to point to his own hand, but not to a pictorial representation of his hand, whereas the patient JO fails to accurately localize her sensations when asked to point to a pictorial representation of her hand, but not to her own hand (Anema et al., 2009).

(iii) During Alice in Wonderland hallucinations, one perceives the size of one's limbs distorted but it has no impact on one's actions (Lippman, 1952).

These dissociations would be difficult to account for if we only had a unique reference frame based on the practical photograph of our body. The only way out is to claim that they are pathological cases and that the practical hypothesis does not apply to them. It would be only under normal circumstances that we use bodily PPRs for localizing our

---

[9] Hence, the practical hypothesis fails to explain the felt location of pain, although at first sight it seems to be one of the most plausible cases. Indeed it cannot account for RB's behaviour. One may further note that the disruption of the practical knowledge of the body (as in ideomotor apraxia) should result in difficulty in localizing painful sensations, but I know of no such report in the literature.

sensations. However, as I shall now argue, even in normal situations bodily PPRs cannot constitute the unique reference frame for bodily sensations. More specifically, I will show that because bodily PPRs guide action they have a specific spatial organization that is incompatible with the spatial focality of bodily experiences: the segmentation of the body into parts is not identical for bodily experiences and for action.

8.4.2 Body Mereology

The body is made up of parts, but how should we differentiate the relevant parts of the body from each other? What form should a mereology of the body take (Vignemont et al., 2005b, 2009)? Should the parts of the body be distinguished from each other on spatial grounds? Functional grounds? Or should we appeal to some other kinds of grounds entirely in carving the body up into parts?

Bermúdez (1998) gives a primary role to what he calls "hinges" for the individuation of body parts. It is actually a well-known effect that tactile localization is improved close to anatomical landmarks that are used as reference points like joints (Weber, 1826; Le Cornu Knight et al., 2014). As joints have a special significance for action by allowing relative movements of body segments, one may then take these results as support for the practical hypothesis. But if this were the right interpretation of the results, then the active use of the joints should reinforce their structuring role. However, we have found the opposite effect (Vignemont et al., 2009). When one remains still, the gap between two tactile stimuli are experienced as wider if the stimuli are applied to two body parts that straddle a joint (e.g. the wrist) than if they are applied within a single body part (within the hand or the forearm). But the effect is *reduced*, instead of increased, when one moves one's hand by rotating one's wrist several times just before being touched. Action appears to diminish the relative distortion of tactile distances across the wrist joint. Other experimental findings using the RHI show that action requires a more holistic representation of the body than purely passive bodily experiences. Tsakiris et al. (2006) compared the scope of the proprioceptive drift after a classic RHI (induced by index fingers stroking) and after an active version of the RHI (induced by voluntary movement of the index finger while watching the index finger of the rubber hand moving in a congruent manner). They found that in the classic RHI, the proprioceptive drift was limited to the index finger, which was stroked, and did not extend to the little finger, which was not stroked. The effect was thus local. However, after active RHI, the proprioceptive drift affected the little finger too, showing a more holistic effect over the whole hand.[10]

Actions thus impose a functional organization because of the sets of body parts that work together in movements. They require an integrative representation of the body that brings effectors together into functional units. This is actually revealed in kinaesthetic experiences: when I feel my hand raising, I actually feel the whole upper limb moving. As Gallagher (1995) claims, "The body schema [...] functions *in a holistic way*" (p. 229, my emphasis). This holistic level of segmentation of the body into parts is indeed the most relevant one for action. It can already be found at the level of the primary motor cortex (M1), by contrast to the primary somatosensory area (SI). Both SI and M1 are organized in a somatotopic manner, thus resulting in the famous homunculus. What is interesting is that the cortical map in M1 is not as well segregated and segmented as that in SI. In particular, the motor cortex is organized in terms of action synergies and patterns of movements (Lemon, 1988; Graziano, 2016). Accordingly, M1 is organized for representing muscle groups that form functionally coherent units rather than individual muscles (Hlustik et al., 2001).

By contrast, the felt location of bodily experiences is generally relatively focal. For example, when I feel touch on my hand, I feel touch just on my hand: my sensation is about my hand, and not about the adjacent body parts like my forearm. Or if I cut myself with a sheet of paper, I feel pain just at the tip of my right index finger. In this sense, bodily experiences can be said to be isolated from the rest of the body. Body mereology thus differs between bodily experiences and actions. Consequently, the practical knowledge of the body is inadequate to account for the focal localization of bodily experiences. It is true that joints are used as bodily landmarks for their felt location, but bodily PPRs carry information about structural affordances that are beyond joint segmentation.

---

[10] One may also interpret the conflicting results about the sensitivity of action to the RHI along these lines. In a nutshell, in a classic RHI, action is immune to proprioceptive drift (Kammers et al., 2009a). But the seen grip aperture of the rubber hand can affect the width of grip aperture of the biological hand (Kammers et al., 2010). The difference in this last study is that both index finger and thumb were stroked together. In other words, the functional unit for grasping was stimulated. Because of this more holistic activation, the RHI modified bodily affordances, and thus had consequences for subsequent actions.

Consequently, they cannot suffice to account for the spatial content of bodily experiences. The practical hypothesis is thus false. Since the body map that is used for action cannot be sufficient for localizing bodily experiences, there must be a different type of body map that is exploited, a well-segmented map that represents categorical body parts, which is of little use for action (Le Cornu Knight et al., 2014). I suggest that this body map is purely descriptive.

I have argued that it makes a difference for the way the structural properties of the body are represented, whether it is for the localization of bodily sensations or for action planning. Put another way, there is not a unique body map but two of them: one that I qualify as being cold and the other as being hot. The *cold* body map is a purely descriptive representation of the enduring properties of the body, which is used for the fine-grained localization of sensations and which has "no inherent activity-valence (it does not dispose us—has no intrinsic connection to a disposition—to being drawn towards or away), and no inherent motivational force" (Cussins, 2012, p. 24). The *hot* body map is both a descriptive and a directive representation of the same properties, whose content itself "disposes the system to intervene actively in the environment" (Cussins, 2012, p. 24). The hot body map cannot suffice to account for the spatial content of bodily experiences, and in particular for their focal localization (e.g. I feel pain at my finger tip). Yet I do not want to fully discard it: bodily experiences are not completely disconnected from action. Instead, they can constitute both justifying and motivating reasons for behaviour, and more particularly for defensive behaviour (Bain, 2013). One may then suggest a duplex characterization of the spatiality of bodily experiences along the visual model proposed by Clark (2001, p. 514). On his view, our visual experience involves "both a relatively passive perceptual content (the perceived filling out of visual space) and the way our visual experience presents that space as an arena for fluent, engaged action".[11] Likewise I suggest that bodily experiences involve both a relatively passive and local spatial content, which is structured by the cold body map, and a more active and holistic content that presents the bodily space as an arena for actions, which is structured by the hot body map.

---

[11] Other versions of the duplex model have been proposed for visual experiences. See for instance Briscoe (2009) and Goodale and Westwood (2004).

8.5 Conclusion

The relationship between perceptual experiences and action has given rise to many philosophical and empirical debates. On the one hand, one may question to what extent and in what manner perceptual experiences contribute to action. On the other hand, one may question to what extent and in what manner action contributes to perceptual experiences. These issues have become even more difficult since the discovery of empirical dissociations between perception and action. These findings do not show that there is no interaction between perceptual experiences and action. They rather invite us to offer a more refined view of their relation. The objective of this chapter was two-fold. Firstly, I have offered a critical analysis of the notions of body schema and body image as they are used in the philosophical and psychological literature. Instead of these unclear notions, I have proposed distinguishing different types of body representations on the basis of their direction of fit. In particular, I have contrasted representations that are purely descriptive with representations that are both descriptive and directive (i.e. bodily PPRs). Secondly, I have analysed the role of bodily PPRs for action and for bodily experiences. In particular, I have argued against the practical hypothesis, according to which the spatial content of bodily experiences can be fully explained by the hot body map. I have defended the view that bodily PPRs are too holistic for the focal localization of bodily experiences. There must be a distinct type of body map, which is cold and which represents body parts along a somatosensory mereology. Those two types of body map, the hot one and the cold one, constitute the reference frames of bodily experiences.

PART III

Bodyguard

Sometimes ... they threaten you with something—something you can't stand up to, can't even think about. And then you say, "Don't do it to me, do it to somebody else, do it to so-and-so ..." At the time when it happens you do mean it. [...] You don't give a damn what they suffer. All you care about is yourself.

(Orwell, 1948, p. 235)

We can eventually turn to the original questions that motivated this book. What is required for the sense of bodily ownership? What distinguishes a body part that is felt as one's own from a body part that is felt as alien? The aim of Part III is to offer the beginning of an answer to these questions. It may not be easy to find it given the numerous differences among the various cases of ownership and disownership, but it goes beyond the scope of this book to systematically analyse the idiosyncrasies of each case and it is more interesting to focus on what they have in common in order to shed light on what is constitutive of the sense of bodily ownership.

In Part I, I highlighted the limits of Martin's spatial conception, according to which the sense of bodily ownership consists in the sense of the spatial boundaries of one's body: it is not sufficient to feel sensations as being located in a body part to experience this body part as one's own. I will now evaluate whether the sense of bodily boundaries is more successful in its account of the sense of bodily ownership if it is enriched with agency, that is, if the sense of bodily ownership consists in the sense of the spatial boundaries of one's body *as being under direct control*. We shall see, however, that this agentive conception cannot account for the specific relationship that holds between bodily control and ownership. I will then further qualify it by introducing an affective dimension. According to the affective conception, the sense of bodily ownership consists in the sense of the spatial boundaries of one's body *as having a special significance for the self*. To defend this view, in Chapter 9, I will analyse the sensorimotor underpinning of the sense of bodily ownership. I will argue that there is a specific type of hot body map that represents the body that has a special evolutionary significance for the organism's needs. In Chapter 10, I will show how this body map gives rise to the phenomenology of bodily ownership, which should be conceived of as an affective tonality of bodily experiences.

9

The Bodyguard Hypothesis

When I report, "I am raising my arm", there are two occurrences of the first person pronoun: at the level of the agentive experience (*I* am raising my arm) and at the level of the body part that is moving (*my arm* is raising). The first expresses the sense of agency. The second expresses a sense of bodily ownership. The question that I want to raise here is how one should understand the relationship between these two types of self-awareness. The point is not to reduce the latter to the former: one can be aware of one's body as one's own while one remains still or during passive movements for which one experiences no sense of agency (Gallagher, 2000; Synofzik et al., 2008). Still there may be a sense in which bodily control, which is at the core of the sense of agency, also contributes to the sense of bodily ownership. More specifically, in 2007 I proposed that the body schema grounds the sense of bodily ownership, and that its disruption causes disownership syndromes (Vignemont, 2007). Since then, I have realized that this agentive hypothesis faces a number of difficulties that cannot be solved without further refinements. In particular, in Chapter 8 we saw that the notion of body schema was too obscure to be of any help and that we should rather use the notion of hot body map. I will now show that even this new notion can be prone to confusion if one does not distinguish between two kinds of hot body maps: a working one involved in instrumental actions, and a protective one involved in self-defence. I shall then argue that only the protective body map can ground the sense of bodily ownership.

9.1   An Agentive Mark for the Sense of Bodily Ownership?

> We posit that the potentiality for action of our bodily self is a necessary condition to accomplish the sense of body ownership we normally entertain.
>
> (Gallese and Sinigaglia, 2010, p. 751)

> The body is a unity of actions, and if a part of the body is split off from action, it becomes 'alien' and not felt as part of the body.
>
> (Sacks, 1984, p. 166)

Intuitively, it might seem that the origin of the sense of bodily (dis)ownership is to be found in bodily control and in its disruption (Davies et al., 2001; Dieguez and Annoni, 2013; Gallese and Sinigaglia, 2010; Rahmanovic et al., 2012; Vignemont, 2007; Baier and Karnath, 2008; Burin et al., 2015; Ma and Hommel, 2015a; Della Gatta et al., 2016 among others). More precisely, it is our *ability* to control our limbs, instead of actual control, that would determine whether we experience them as our own or not. But is it a necessary condition for the sense of bodily ownership? Does one experience a body part as one's own only if one can control it? And is it a sufficient condition? Does our ability to control a body part entail that one experiences it as one's own?

9.1.1   The Body Under Control

Consider first the case of somatoparaphrenia, in which the relationship between action and the sense of bodily ownership is especially salient: in almost no reported cases (with the exception of two cases, see Vallar and Ronchi, 2009) can somatoparaphrenic patients control their "alien" limb. Most of the time, they are paralysed, and when they are not, they suffer from the anarchic hand sign (i.e. the limb is experienced as having a will of its own). Somatoparaphrenic patients frequently complain about the uselessness of their "alien" limb, which is said to be good for nothing, lazy, or worthless, like a sack of coal (Feinberg, 2009). One may then be tempted to explain their sense of disownership as follows:

> The patient with somatoparaphrenia is no longer able to move her paralysed limb, which is at odds with her prior experience of her limb. This generates the thought that the limb cannot be hers: it is an alien limb. This initial thought is then accepted uncritically as true.   (Rahmanovic et al., 2012, p. 43)

To test this hypothesis, one can also look at non-somatoparaphrenic patients who have motor deficits (because of schizophrenia, focal hand dystonia, or spinal cord injury) and analyse how they experience their limbs. It has been argued that their sense of bodily ownership is diminished, and that they are thus more susceptible to appropriate another limb and more prone to experience the RHI (Burin et al., 2015; Fiorio et al., 2011; Peled et al., 2013; Scandola et al., 2014, 2017). A further potentially interesting disorder, is ideomotor apraxia insofar as apraxic patients are said to have a deficit of body schema (Buxbaum et al., 2000). Their sense of their body is rarely explicitly investigated but it is interesting to note that the patients report that they can "lose" their hands: "All the participants experienced to some degree that their hands had to be searched for, found, put in the right place and 'monitored' through the action" (Arntzen and Elstad, 2013, p. 69).

Whereas the loss of bodily control seems to induce a sense of dis-ownership, its recovery can re-establish a sense of ownership. This is the case in deafferented patients. We saw earlier that Ian Waterman reported first feeling alienated from his body but as he regained control over his body he also regained a sense of bodily ownership (Gallagher and Cole, 1995). Likewise, amputees who have received hand transplants experience their new hands as their own only when they can control them (after eighteen months), and not earlier when they gain sensations in them (after six months) (Farnè et al., 2002):

DC (who received a bilateral hand allograft): At the beginning I felt nothing, and then, after six months after the transplantation, incredible: I felt pain, hot, cold, light touch, textures [...] Journalist: Was it difficult to accept life with someone else's hands? DC: Yes, it took more than a year. I used to say "the hands", but not "my hands" [...] Journalist: When did you really appropriate these hands? DC: When the first phalanxes started to move, when I was able to eat again with a fork. Then, I said: "Here are my hands." Now I'm completely normal.   (Interview of DC, 2008)[1]

The importance of control is finally confirmed by illusions of ownership in a virtual reality environment: controlling a virtual avatar can make one experience the avatar's body as one's own (e.g. Slater et al., 2008).

At this point, one may simply reply that the actual capacity for bodily control cannot be necessary for the sense of bodily ownership given the number of individuals who are paralysed and who still experience their body as their own. Indeed disownership syndromes are very rare whereas paralysis is not, and many patients who are paralysed do not feel disownership. Hence, the loss of control cannot be sufficient for the

---

[1] "L'important: jouer avec mes enfants", *Le Matin*, 26 August 2008, my translation.

loss of the sense of bodily ownership. An alternative view may then appeal to *agentive feelings*: the body that I experience as my own is the body that I feel that I can control. However, this proposal is both too conservative and too liberal. It is too conservative because it cannot account for cases of individuals who feel that they have no control over their limbs and still experience them as their own, such as schizophrenic patients with delusion of control and patients with "pure" anarchic hand: they feel that their limbs do not obey their will but they still feel that their limbs belong to them. Hence, it is not necessary to feel that one can move a limb in order to experience it as one's own. On the other hand, this proposal is too liberal because the fact that one feels that one can move a limb does not ensure that one experiences it as one's own. This is clearly demonstrated by patients with somatoparaphrenia, who can be unaware of their paralysis (i.e. anosognosia for hemiplegia): they erroneously feel that they can control their paralysed "alien" hand, and yet they do not experience it as their own.

Let us thus turn to a further version of the agentive conception, which does not appeal to agentive feelings but only to *sensorimotor representations*: the parts of the body that I experience as my own are those that are included in the hot body map. The hot body map represents the body that is under control but it can also misrepresent it. It can thus be dissociated from actual motor capacity. Consequently, the hot body map can still include legs that are actually paralysed. It can also be dissociated from the conscious experience that one has of one's motor abilities. In anosognosia for hemiplegia, for example, the sensorimotor representation of the paralysed side of the body is disrupted, and yet patients still feel that they can move it. This sensorimotor proposal can thus avoid what appeared as fatal objections to other agentive proposals. It assumes that it is both necessary and sufficient for a limb to be represented in the hot body map for one to experience it as one's own. However, we shall see that this sensorimotor proposal is also both too liberal and too conservative.

9.1.2 The Body That One Plans One's Actions With

If the hot body map grounds the sense of bodily ownership, then action planning, which is based on the hot body map, should be modified by the appropriation of extraneous body parts and by the exclusion of one's body parts. If the agentive conception is true, one should thus expect specific motor correlates to the senses of ownership and of disownership. Let us first consider the sense of disownership. According to the agentive conception, it should be accompanied by the inability to correctly control the limb that has been excluded from the hot body map. Such a prediction is confirmed by a study using hypnosis (Rahmanovic et al., 2012). In this study, one group received a suggestion about disownership: "Whenever I tap my pen like this [tap pen three times on table], this arm [touch targeted arm] will feel that it belongs to someone else". This suggestion successfully induced disownership. Three participants even maintained their experience in the face of contradiction. For instance, a participant reported after the disownership suggestion:

Wasn't feeling like the rest of my body ... This [targeted] hand was like it was a dead weight ... it was just there, but it wasn't part of me ... and there wasn't any awareness that it was mine.   (Rahmanovic et al., 2012, p. 54)

Interestingly, the disownership suggestion also induced as much paralysis as the paralysis suggestion that was given to a second group of participants ("Whenever I tap my pen like this [tap pen three times on table], this arm [touch targeted arm] will feel paralysed").[2] Hence, not only can paralysis cause an experience of disownership (as in somatoparaphrenia) but it also can be its collateral damage. As far as disownership is concerned, the agentive conception thus seems to be confirmed.

Let us now consider the motor correlates of the sense of bodily ownership. There are situations in which action planning is modified by the incorporation of extraneous body parts. This is the case in some versions of the RHI, as seen in Chapter 8 (e.g. Kammers et al., 2010). This is also the case in the embodiment delusion. In this delusion, patients report a sense of ownership towards another person's arm when they see it in front of them next to their own left arm. Interestingly, although they are paralysed, they report feeling as if they were moving when they see the "embodied" arm moving. Even more strikingly, it has been found that they process the movements of the "embodied" hand as their own. In one study, they were asked to draw a circle with their right hand

---

[2] With the paralysis suggestion, some members behaved as if they could no longer move but they did not report disownership experiences. This confirms what we already know from the paralysed patients who continue to experience their body as their own.

while observing the "embodied" hand drawing a vertical line. Their performance revealed a classic interference effect that normally occurs only when the task is performed by the subject's own two hands: the patients drew trajectories assuming oval shapes instead of circular ones (Garbarini et al., 2012; see also Garbarini et al., 2015). One way to interpret this result is that the "embodied" hand has been included in the hot body map. Again, this is in line with the agentive conception.

However, there are other situations in which action planning seems to be impervious to the embodiment of extraneous hands. We have seen that in the RHI the motor system does not always take the location of the rubber hand as a starting parameter when planning reaching and grasping movements (Kammers et al., 2009a). A critic of the agentive conception may then argue that this result suffices to show that the hot body map does not ground the sense of bodily ownership. The objection runs as follows:

(i) In this study motor immunity to the illusion shows that the rubber hand is left out of the hot body map that guides the movements;

(ii) Yet participants report ownership over it;

(iii) Thus, it is false that one experiences as one's own only the body parts that are represented in the hot body map.

Another objection against the agentive conception derives from tool use. We saw that tools can be motorically embodied to such an extent that after tool use, one programmes one's movements as if one were still holding the tool, even for actions that have never been performed with it (Cardinali et al., 2009a). These findings indicate that the hot body map is updated during tool use. The critical question is why one generally experiences no ownership for tools. The objection thus runs as follows:

(i) Tools are included in the hot body map that guides reaching and pointing movements;

(ii) Yet participants can report no ownership over them;

(iii) Thus, it is false that one experiences as one's own any body parts that are represented in the hot body map.

One might be tempted to reply that the hot body map can make us feel an object as part of our body only if the object looks like a body part (de Preester and Tsakiris, 2009). In other words, if tools are not experienced as parts of our body, it is because of their non-bodily shape. Some studies have indeed shown that one cannot induce the RHI for a wooden object that does not look like a hand, such as a rectangular block (Tsakiris et al., 2010). However, a recent study has been able to induce an illusion of ownership for a tool (Cardinali et al., under revision; see also Ma and Hommel, 2015b). Interestingly, the actual use of the tool played no role in this study: it was induced by the classic RHI paradigm: both the participant's hand and the tool, which participants had never used, were synchronously stroked. Even after participants used the tool, its control did not have an impact on their rating of ownership.[3] On the other hand, even when tools are completely hand-shaped, as is the case for prostheses, they can be difficult to appropriate. As mentioned in Chapter 2, most amputees feel their prostheses as completely extraneous. For example, a patient reported:

Using a prosthetic is not a natural thing, because a prosthetic is not a substitute leg, *it is a tool* which may or may not do some of the things that a leg might have done. (Murray, 2004, p. 971, my emphasis)

To recapitulate, the agentive proposal faces two objections: (i) one can experience the rubber hand as part of one's body although it does not seem to be incorporated in the hot body map; and (ii) one does not experience tools as parts of one's body although they seem to be represented in the hot body map. Do these objections entail that the sense of bodily ownership bears no relationship whatsoever with bodily control? I do not think so. I will now argue that there is an equivocation of the term "hot body map", which can refer to two distinct types of PPRs. More specifically, I propose that one must distinguish what I call the *working body map*, which is involved in instrumental movements, from the *protective body map*, which is involved in self-defence.

9.1.3 The Duality of Hot Body Maps

Let us consider again the RHI. It is true that one can experience a rubber hand as part of one's body while failing to plan one's movements on the basis of its location. However, the movements that are tested with the

---

[3] Along the same lines, it has been shown that the involvement of action does not make the RHI stronger: the ownership rating for the rubber hand is not significantly different whether the illusion is induced actively (seeing the rubber hand moving in synchrony with active movements of the biological hand) or perceptually (seeing the rubber hand touched in synchrony with tactile stimulation on the biological hand) (Kalckert and Ehrsson, 2014).

RHI are goal-directed instrumental movements, such as pointing and grasping, and there is a different range of movements that is worth exploring, namely *defensive movements*. Physiological response to threat (measured by skin conductance response, or SCR) has indeed become the main implicit measure of the RHI:[4] it has been shown repeatedly that participants react when the rubber hand is threatened, but only when they report it as their own after synchronous stroking, and the strength of their reaction is correlated with their ownership rating in questionnaires (for the RHI, see Armel and Ramachandran, 2003; Ehrsson et al., 2007; Riemer et al., 2015; Zhang and Hommel, 2015; for full-body illusions, see Hagni et al., 2008; for the tool ownership illusion, see Cardinali et al., under revision). In the embodiment delusion too, when patients see a needle approaching their "embodied" hand, they show an increase in SCR (Pia et al., 2013; Garbarini et al., 2014; Fossataro et al., 2016).

One may question how to precisely interpret this measure. It can indeed express a general level of arousal, which is not specifically related to the sense of bodily ownership (Ma and Hommel, 2013). However, it can also express anxiety and further measures have confirmed that this is the case in the RHI: the sense of ownership induced by synchronous stroking is associated with brain activity commonly found for subjective anxiety in the insular and anterior cingulate cortex (Ehrsson et al., 2007; Zhang and Hommel, 2015). One might then be tempted to interpret these results in terms of empathy or contagion. However, one should not too hastily assimilate our reaction to danger when it is directed towards us and towards someone else. In the studies mentioned here, there is always a significant physiological difference when participants experience the rubber hand or another person's hand as their own and when they do not. For instance, in the study on the embodiment delusion, the effect is specific to the other person's *left* hand (the one that the patients experience as their own) and when the experimenter hurts the other person's *right* hand (for which they have no sense of ownership), there is a significant difference. Moreover, control participants show no similar

---

[4] This is actually an old measure used to test phantom limbs, which is known as the Abbattucci's *choc à blanc*: when amputees see that the location at which they feel their phantom limb to be hit, or merely threatened, they show extreme distress, or even pain, and immediately withdraw their phantom limbs, as shown by the movements of their stump (which was never in danger).

significant increase in SCR when they see the other person's hand being injured (Pia et al., 2013).

Hence, when there is an increase in SCR, it does not reveal a kind of empathy or contagion but rather a protective response that one normally has for one's own body, which can also be revealed at the motor level. In one study, participants were in an immersive virtual reality system, which induced them to feel ownership over a virtual hand. The virtual hand was then attacked with a knife. Although participants were instructed not to move, their event-related brain potentials revealed that the more they felt the virtual hand as their own, the more the perception of threat induced a response in the motor cortex (Gonzalez-Franco et al., 2013). Roughly speaking, when one observes the rubber hand that one experiences as one's own being threatened, one automatically withdraws one's hand. This finding suggests that the RHI has a specific agentive mark in the context of self-protection.

Let us now reconsider the objection from tool use. We saw in Chapter 2 that Butler (1872, p. 267) claimed, "We do not use our own limbs other than as machines", but contrary to what he assumed we do use our own limbs other than as machines. More specifically, we do protect them other than as machines. This hypothesis has been tested by Povinelli and colleagues (2010) on chimpanzees. Chimpanzees had the choice between using their hand or a tool to open a box. It was found that they removed the cover of the box with their hand when they perceived that it contained food and with the tool when they perceived the object in the box as potentially dangerous: "the target of the actions may be located well within reach, but a tool is chosen as a substitute for the upper limb in order to avoid harm" (Povinelli et al., 2010, p 243).

More generally, we use tools in harmful situations in which we would not use our limbs and we easily put them at risk as long as there is no risk for our body. If we protected tools as we protect our limbs we would not be able to use them as extensively as we do, and the range of our actions would be far more limited: we could no longer stoke the hot embers of a fire or stir a pot of boiling soup. This claim does not contradict the fact that we protect tools and react if they are under threat (Rossetti et al., 2015). We actually need to keep them in good shape in order to be able to use them. Nonetheless, their significance is not of the same kind as the significance of our own body and it is likely that their protection is subserved by different mechanisms than bodily protection.

We thus have a double dissociation: the rubber hand can be incorporated for planning protective movements but not instrumental ones, whereas tools can be incorporated for planning instrumental movements but not protective ones. I propose that there are two kinds of hot body maps that underlie these distinct types of actions. The *working body map* consists in bodily PPRs used for acting on the world, for grasping, exploring, manipulating, and so forth. Most literature in neuropsychology and in cognitive psychology has focused its interest exclusively on this notion (under the label of body schema), but there is another type of bodily PPR that one should not neglect, which is involved in avoiding predators and obstacles, namely the *protective body map*. The relatively uncontroversial starting point is the hypothesis that the body matters for survival and needs to be defended:

By far the chief pre-occupation of wild animals at liberty is finding safety, i.e. perpetual safety from enemies, and avoiding enemies. The be-all and end-all of its existence is flight. Hunger and love occupy only a secondary place, since the satisfaction of both physical and sexual wants can be postponed while flight from the approach of a dangerous enemy cannot. In freedom, an animal subordinates everything to flight; that is the prime duty of an individual, for its own preservation, and for that of the race.   (Hediger, 1950, p. 20)

Because of the significance of the body for the organism's survival, Hediger proposes that animals do not process space uniformly and in particular that there is a specific zone immediately surrounding their body, described as the flight distance, that predators cannot approach without eliciting specific responses (flight or fight depending on how close the predator is). Because of the significance of the body for the organism's survival, I propose that there is also a specific representation to fix what is to be protected. One does not protect one's biological body; one protects the body that one takes oneself to have and the protective body map, like any representation, can misrepresent one's biological body and include a phantom hand, a rubber hand, or another person's hand. Although rarely mentioned in the literature on body representations, this notion of body map can be found in the literature on pain (Klein, 2015a; Moseley et al., 2012a):

There's a body schema representation which is primarily concerned with protective action: that is, one which maps out parts of our bodies that we should pay special attention to, avoid using, keep from contacting things, and so on. Call this a defensive representation of the body: it shows which parts of the body are in need of which sorts of defence.   (Klein, 2015a, p. 94)

9.2 A Dual Model of Peripersonal Space

I have argued that the protective body map has the same origin as the flight zone: self-preservation. One may then be able to further explore this specific type of body representation, and its difference with the working body map, by considering in more detail how we represent the space immediately surrounding our body. We shall see that peripersonal space can be represented in two different ways depending on its function, whether it plays a role for instrumental or for protective movements. Each notion of peripersonal space in turn requires distinct body maps.

9.2.1 A Body-Centred Space

The defining characteristic of peripersonal space is that it is encoded in a body part-centred frame of reference. It is anchored in specific parts of the body and when they move, what is represented as peripersonal space also moves. In Chapter 5, we saw that the representation of peripersonal space is multisensory so that visual or auditory stimuli presented close to a specific limb interfere with the processing of tactile stimuli applied to this limb (i.e. cross-modal congruency effect). For instance, when the hands are uncrossed, visual stimuli close to the left hand presented on the left side of the body affect tactile processing on the left hand. When the hands are crossed, visual stimuli presented at the same egocentric location (on the left) but now close to the right hand affect tactile processing on the right hand (Figure 9.1). What matters is bodily location (left hand versus right hand), and not egocentric location (on the left or on the right).

Not only does peripersonal space follow the body part that anchors it, it can also stretch and shrink as the estimated size of the body part stretches and shrinks (Maravita et al., 2002; Holmes et al., 2007; Bassolino et al., 2010; Canzoneri et al., 2013b). In a seminal study Iriki and colleagues (1996) trained monkeys to use a rake to reach food placed outside their reaching space. They found that some neurons, which displayed no visual response to food at this far location before training, began to display visual responses after training (Figure 9.1). A few minutes after tool use was interrupted,

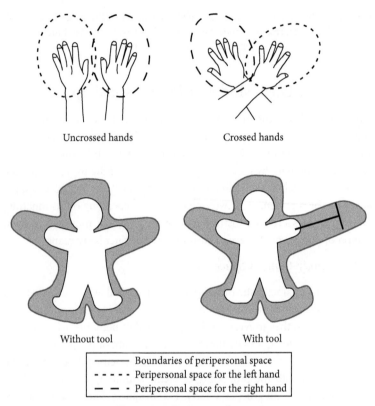

Uncrossed hands   Crossed hands

Without tool   With tool

	Boundaries of peripersonal space
- - - - -	Peripersonal space for the left hand
— — -	Peripersonal space for the right hand

Figure 9.1. Peripersonal space.

the visual receptive fields shrank back to their original size.[5] Another interesting example of plasticity follows from the use of prosthesis (Canzoneri et al., 2013a). When amputees wear their prostheses, their peripersonal space stretches farther, and the variation of the extension of their peripersonal space follows the variation of their perceived arm size (they perceive their arm as longer with prosthesis than without it).

Hence, peripersonal space surrounds bodily space as it is represented in the body map, and more specifically in the *hot* body map. As Rizzolatti et al. (1997) describe, peripersonal space has indeed a "praxic" function:

---

[5] Furthermore, it has been shown that visual stimuli presented in far space induce more severe extinction immediately after tool use, compared with before tool use (e.g. Berti and Frassinetti 2000; Farnè and Ladavas 2000; Maravita and Iriki 2004).

it organizes sequences of movements. There are, however, two ways to understand its praxic function: it can be for acting on the environment or for protecting the body from it.

9.2.2 Working Space and Protective Space

Peripersonal space is often analysed in terms of reaching space and affordances (e.g. Coello et al., 2008). Whether these notions refer to the same type of phenomenon or not (Cardinali et al., 2009b), it is clear that peripersonal space is a privileged interface for acting on nearby objects (Serino et al., 2009). In monkeys, merely seeing graspable objects activates specific neurons in the motor cortex (known as canonical neurons) (Jeannerod et al., 1995). In humans, automatic motor activations are described during the observation of manipulable objects (Chao and Martin, 2000; Grafton et al., 1997). In addition, there is evidence of stimulus-response compatibility effects that occur only if the stimulus is presented in peripersonal space (e.g. Tucker and Ellis, 1998; Costantini et al., 2010).[6]

As described by Hediger (1950), peripersonal can also be conceived of as a zone of safety, which requires appropriate actions if it is invaded. Electrophysiological studies on monkeys have confirmed that stimuli close to the body can elicit avoidance responses (Graziano and Gross, 1993). Likewise, we lean away from a visible object, and, when walking through a doorway we tilt our shoulders to protect our body from hitting the doorframe. Other defensive behaviours require no action at all, like freezing or playing dead. For example, an intense sound near the hand can cause a defensive-like freeze response in humans (Avenanti et al., 2012), resembling that observed during the presentation of noxious stimuli (Cantello et al., 2000). What the protective function of peripersonal space confirms is the fact that the body has a special significance for the organism's evolutionary needs and that because of it, the brain dedicates specific resources to represent its immediate surrounding.

On the basis of this functional difference, we suggest that there are at least two ways of representing peripersonal space: the *working space* for

---

[6] Here is an example of this effect. Participants are instructed to press a button with their left or right hand depending on whether the object is upright or inverted. The motor response is facilitated if the graspable part of the object (e.g. the handle of a saucepan) is aligned with the hand that responds, and this is so only when the object is within peripersonal space (Constantini et al., 2010).

taking advantage of opportunities (e.g. to grasp food and useful objects) and the *protective space* for protecting the body from potential threats (e.g. to avoid a bee flying towards one's face) (Vignemont and Iannetti, 2015). Difference in function between the two types of peripersonal space has a strong impact at the sensory level because we do not need to extract the same type of information about an object that we want to avoid and that we want to grasp. The protective function primarily requires rapid detection of what can be conceived of as a threat. By contrast, the exploratory function primarily requires recognition of the fine-grained features of the object (its 3D shape and its length, for example), to precisely guide the movement towards it. Furthermore, one can isolate two distinct neural circuits (Clery et al., 2015). The circuit of protective space is formed by parietal area VIP and premotor area F4 (Rizzolatti and Luppino, 2001; Matelli and Luppino, 2001):

this network is proposed to subserve the representation and protection of near peripersonal space or safety margin around the body, with a specific emphasis on two vulnerable body parts, the head and the arm/hand unit.

(Clery et al., 2015, p. 319)

By contrast, the circuit of the working space is formed by parietal areas AIP and 7b and premotor area F5:

It is functionally specialized in the visuomotor transformation that subserves the grasping of objects in our environment, i.e. the online adjustment of the hand and finger configuration for a secured interaction with the objects.

(Clery et al., 2015, p. 320)

If there are indeed two distinct neural circuits, associated with two distinct types of sensorimotor processing of peripersonal space, then it should be possible for their operations to be differentially modulated.[7] Interestingly, anxiety seems to play just this role. Anxious individuals *underestimate* the extension of their working space because they under-estimate their capabilities (Graydon et al., 2012; Lourenco and Longo, 2009; Bassolino et al., 2015), but they *overestimate* the extension of their protective space, either because they feel less able to react to threats or because they overestimate potential danger (Lourenco et al., 2011; Sambo

---

[7] The difficulty in finding dissociations is that the main experimental design to test peripersonal space is the cross-modal congruency effect, which measures both types of peripersonal space.

and Iannetti, 2013). The data derived from studies on tool use also points to dissociation between working space and protective space. As mentioned earlier, there are a large number of results showing that tool use extends peripersonal space further away from the subject, but these studies involve performing goal-directed actions such as reaching and grasping with the tool, and thus show that tool use stretches the *working space* only (Iriki et al., 1996; Maravita et al., 2002; Maravita and Iriki 2004; Farnè et al., 2007; Serino et al., 2007).[8]

To each function of peripersonal space thus correspond differences in information processing. Taken together these results argue in favour of a distinction between two representations of peripersonal space, respectively anchored in two distinct types of hot body maps (see Table 9.1). Working space is a privileged interface for the body to act on nearby objects, which is anchored in the working body map. Protective space is a privileged interface for detecting potential threats to avoid, which is anchored in the protective body map.

Table 9.1. A dual model of peripersonal space

	Protective space	Working space
Function	To protect one's body	To take advantage of opportunities
What is salient	Potential threats	Affordances
Prototypical movement	Withdrawal	Grasping
Bodily anchor	Protective body map	Working body map
Sensory priority	Fast detection	Fine-grained processing of spatial properties
Spatial priority	Face and hands	Hands
Plasticity	Anxiety: expansion	Anxiety: reduction
	Tool: no expansion	Tool: expansion

[8] One study reported that after using a 45-cm-long tool participants reacted more to threats occurring at 40 cm from their body than before using it (Rossetti et al., 2015). However, one cannot conclude that the space surrounding the tool is processed as protective space. One would indeed need to test affective response to a threat occurring 40 cm *away* from the end of the tool itself.

9.3 Fixing the Boundaries of the Body to Protect

Now that we have a clear distinction between the protective and the working body maps, we can narrow down the agentive conception and propose the following hypothesis, which I call the bodyguard hypothesis:

> *The bodyguard hypothesis:*   One experiences as one's own any body parts that are incorporated in the protective body map.

Given the suite of cognitive capacities that human beings normally have, the protective body map grounds the sense of bodily ownership.[9] The bodyguard hypothesis falls in line with the general agentive conception because it posits a hot body map at the core of the sense of ownership, but it includes a new dimension, namely an affective one, because the protective body map represents the body that has a special value for the organism's evolutionary needs. This specific agentive proposal can avoid the objections that I raised earlier on the basis of the RHI and of tool use: the rubber hand is incorporated into the protective body map and thus it is felt as of one's own, whereas the tool is usually not and thus it is not felt as of one's own. We shall now see that the bodyguard hypothesis can also account for the sense of bodily disownership.

9.3.1 The Affective Mark of Bodily Disownership

How does one react when one experiences as alien the body that is under threat? So far I have described illusions of bodily ownership but it is also possible to induce illusions of disownership (Newport and Gilpin, 2011; Gentile et al., 2013). For example, in Newport and Gilpin's study, healthy participants place their hands inside a box, which allows them to see live video images of their hands. They are then asked to reach across to touch their right hand with their left hand. In one condition, the experimenters create a situation in which the right hand unexpectedly disappears from vision and touch. All that can be seen and felt is the table where the right hand has once been (unbeknownst to the participants, their right hand has slowly moved outwards). Participants then report that they feel that their right hand is no longer part of their body and they do not react when they see the experimenter pretending to stab the last seen location of their right hand or its real hidden location.

---

[9] Here I limit the scope of the bodyguard hypothesis to humans, leaving aside the delicate issue of the sense of bodily ownership in other animals.

Likewise, patients who suffer from disownership syndromes do not react when their "alien" hand is threatened. In one study on somatoparaphrenia, patients saw either a Q-tip or a syringe approaching either their right hand, which they felt as their own, or their left hand, which they felt as alien (Romano et al., 2014). The experimenter then measured their SCR. When the syringe approached the right hand, the SCR increased, as expected. But when the syringe approached the left "alien" hand, there was no modification of the SCR. These findings are consistent with their broad pattern of attitudes. As described by Oliver Sacks (1984) in his account of his own episode of bodily disownership, "there was a 'sympathetic' deficit, so that I had lost much of my feeling for the leg" (p. 54, my emphasis).[10] Many patients with somatoparaphrenia often try to get rid of their "alien" limb by pulling it out of their bed, giving it to the doctor, putting it in the garbage, and so forth. They can also display misoplegia (i.e. dislike of one's body) and self-inflicted injuries, as illustrated by the following description of a patient's behaviour:

The patient regularly hit the arm on a table violently, scorched it with his right-hand nails and pointed objects such as needles, and sometimes burned it with a cigarette. He also abused the limb verbally, swearing at it and calling it "the damn useless thing." He sometimes contemplated the idea of "chopping it off with an axe".
(Dieguez and Annoni, 2013, p. 178)

In xenomelia, some patients not only contemplate the idea, but actually chop their "alien" limbs off. When tested, it was found that they reacted in the same way as patients with somatoparaphrenia: they showed no increase of SCR when the limb that they no longer wanted was threatened (Romano et al., 2015). In depersonalization too, patients show a lack of affective response when their body is under threat (Dewe et al., 2016). The hypothesis is thus that in those disownership syndromes, the protective body map no longer represents the limb, which thus feels alien.[11]

[10] Interestingly, it has also been proposed that anosognosia for hemiplegia, which is often associated with somatoparaphrenia, should also be conceived of as a kind of detached attitude from one's body: "Detachment would amount to the patient 'withdrawing inhabitation' of the body part and ceasing to care about it as part of their bodily self" (Marcel et al., 2004, p. 36; see also Vuilleumier, 2004; Fotopoulou et al., 2010).
[11] This does not entail that it is the only deficit in somatoparaphrenia. For instance, a somatoparaphrenic patient had difficulty in reconstructing a body and a face by using pre-cut puzzle pieces, thus indicating that his visuo-spatial representation of his body was impaired too (di Vita et al., 2015).

However, one might wonder whether the bodyguard hypothesis is compatible with the fact that patients with somatoparaphrenia and xenomelia can still feel pain as being located in their "alien" limb and negatively react to it. As seen in Chapter 2, one somatoparaphrenic patient asked to have his alien arm removed and put in a cupboard to stop the pain (Maravita, 2008). Since the protective body map no longer includes the limb, one might have expected these patients to behave like depersonalized patients who experience pain as if they were not concerned: "it is as if I don't care, as if it was somebody else's pain" (Sierra, 2009, p. 49). There is, however, a difference between these disorders. In depersonalization, there is a fundamental disruption of subjectivity, which impacts not only their sense of bodily ownership but also their sense of mental ownership (Billon, 2017; Vignemont, 2017c). By contrast, somatoparaphrenia patients can still feel concern for themselves, or more precisely for their feelings. Still their pain behaviours is not exactly normal. In short, to want one's limbs to be dismantled is clearly not a good way to protect one's body. Arguably, the aversive reaction in somatoparaphrenic patients does not recruit the protective body map. In this sense it can be said to be disembodied.

To recapitulate, we react differently whether we experience the body under threat as our own or not. Our affective response thus constitutes a robust marker for the sense of bodily ownership (see Table 9.2). The bodyguard hypothesis explains such a correlation by appealing to a third

Table 9.2. An affective mark of the sense of bodily ownership

	Sense of bodily ownership	Affective response to threats
One's own body under normal conditions	YES	YES
Phantom limb	YES	YES
Rubber hand illusion	YES	YES
Ownership illusion for tools	YES	YES
Embodiment delusion	YES	YES
Somatoparaphrenia	NO	NO
Depersonalization	NO	NO
Xenomelia	NO	NO
Congenital insensitivity to pain	NO	NO
Disownership illusion	NO	NO
Regular tool use	NO	NO

term: the protective body map both motivates one's affective responses to threat and grounds the sense of bodily ownership.

9.3.2 Spatial and Affective Requirements

I have just argued that the boundaries of the body that I experience as my own are those represented in the protective body map, but what determines the content of the protective body map? We have seen that situations in which it includes a new limb or object are quite exceptional. In particular, we have seen that it does not integrate the tools that we use in everyday life. This lack of integration is important because it prevents us losing track of our biological body: thanks to the unaltered protective body map, we keep a default body map that can be used to recalibrate the working body map once we drop the tools. Most of the time the protective body map fulfils its function and targets exclusively one's own biological body. But this is not always the case and we still need to understand how in some rare circumstances it can include extraneous objects or exclude one's own body parts.

We know that it is a representation of the body that is action-orientated, which indicates that it must be calibrated by motor factors, but we also know that it can incorporate a purely passive rubber hand, whereas it fails to do so for a tool under control. Hence, bodily control can hardly be the only factor that decides what is incorporated by the protective body map. More than a hundred versions of the RHI have tried to determine these other factors and the most recent computational models of the RHI suggest that what is included in the protective body map results from the Bayesian weighting of multiple cues of various types (Armel and Ramachandran, 2003; Kammers et al., 2009b; Hohwy and Paton, 2010; Apps and Tsakiris, 2013; Samad et al., 2015). I do not intend here to give a computational model of the protective body map, nor to provide an exhaustive list of the multiple cues and factors that are involved. Instead, I want to highlight the fundamental building blocks for such a model.

To fix the bodily boundaries that have evolutionary significance, I argue, the protective body map must answer two questions: (i) where does the body stop, and the rest of the world start? and (ii) what matters for self-preservation? As we shall see, touch and vision primarily answer the first question, whereas pain primarily answers the second question.

The first question concerns the sense of spatial boundedness. We saw in Chapter 2 that, on Martin's (1992) view, tactile experiences suffice to provide such spatial awareness. Thanks to touch one can actually delineate the apparent limits of the body in two distinct ways: (i) by being aware of the locations at which one can feel tactile sensations and those at which one cannot; and (ii) by being aware of the resistance of the external world on the skin. However, touch is not our only resource for spatial awareness and one should not neglect the role of vision. As argued in Chapter 6, visual information about the body plays a constitutive role in establishing the body. In particular, it provides information about shape: for an object to be integrated in the protective body map, it must look *functionally* identical to a body part (Aymerich-Franch and Ganesh, 2016) and in physical *continuity* with the rest of the body (Tieri et al., 2015b).[12] Vision also provides information about what is immediately beyond one's body. To see peripersonal space is precisely to be aware of the body as a bounded object among other objects within a body-centred frame of reference.[13]

Vision thus provides information about the body and its immediate surrounding, and in combination with touch it fixes its spatial boundaries. However, we should not stop here in our account of the protective body map. It does not simply represent spatial boundaries—it represents spatial boundaries that have an *affective valence*, which is at the origin of its motivational force. To fix these boundaries involves determining which body has a specific value.

It is at this affective level that pain, interoception, and other affectively loaded bodily sensations, can intervene. Neither touch nor vision has intrinsic affective valence: they give no information about the value of the body for the organism.[14] By contrast, pain gives a valence to bodily

---

[12] As argued in Chapter 8, hot body maps represent bodily affordances. Hence, the protective body map can be relatively tolerant regarding fine-grained anatomical appearance as long as the object affords the same type of movements, such as grasping (Ma and Hommel, 2015b; Cardinali et al., under revision). In other words, only functional bodily integrity must be preserved. For example, the sense of ownership for a virtual forearm is significantly reduced when the wrist of a virtual forearm is missing (Tieri et al., 2015a).

[13] Several accounts of the RHI actually ascribe a key role to peripersonal space for the sense of bodily ownership (Blanke et al., 2015; Makin et al., 2008).

[14] The exception is affective touch. Interestingly, the RHI is stronger if it is induced by pleasant stroking (Crucianelli et al., 2013). However, affective touch has been primarily shown to play a role for *social bonding*: the rewarding value of affective touch would reflect

boundaries (Vignemont, 2017b). It indicates that these boundaries are the ones to care about and to protect if one wants to survive. It vividly highlights for the subject that what is inside one's bodily boundaries matters. This does not mean that it is only when the body is in danger that we experience our body as our own. I defend a weaker conception, according to which it suffices that it has been in danger at some point to set up a specific valence to the boundaries of the body. Arguably, there is a stage in development in which it is important, possibly even necessary, to experience pain or other negatively loaded sensations in order to give affective significance to the boundaries of the body, a significance that is at the origin of the sense of bodily ownership. In short, past experiences that the boundaries of the body could be at risk make these boundaries the boundaries of the body that we now experience as our own.

The importance of pain for the sense of bodily ownership is well illustrated by considering the phenomenon of pain deprivation.[15] Patients with congenital insensitivity to pain are characterized by dramatic impairment of pain sensation since birth, caused by a hereditary neuropathy or channelopathy. They show a complete lack of discomfort, grimacing, or withdrawal reaction to prolonged pinpricks, strong pressure, soft tissue pinching, and noxious thermal stimuli. Interestingly, they describe their body feeling like an external object, as a kind of tool. An 18-year-old boy with congenital pain insensitivity reported:

A body is like a car, it can be dented but it pops out again and can be fixed like a car. Someone can get in and use it but the body isn't you, you just inhabit it.

(Frances and Gale, 1975, pp. 116–17)

It thus seems that Descartes (1641) was right: "Nature likewise teaches me by these sensations of pain, hunger, thirst, etc., that I am not only lodged in my body as a pilot in a vessel" (*Meditation VI*). Without some

an evolutionary mechanism that promotes interpersonal touch and affiliative behaviour, which in turn favours the development of a number of cognitive abilities (for a review, see McGlone et al., 2014). The function of affective touch thus seems directed more towards the proximity between bodies than towards their discrimination.

[15] It is also well illustrated by tools. As argued in Chapter 2, one can have referred tactile sensations in tools (one locates in the knife itself the resistance of the carrot that one is cutting, for instance) but one does not have referred painful sensations in them. Consequently, the protective body map does not represent tools, and thus one does not experience them as parts of one's body.

past experiences of pain, I am only as a pilot in a vessel, and this is so even if I still experience other types of bodily sensations. Patients with congenital pain insensitivity can indeed feel touch, and even self-touch, but this does not suffice to compensate for pain deprivation. Touch and vision contribute to the delineation of the boundaries of one's own body, but they do not contribute to the sense of bodily ownership in the way that pain does.

9.4 Conclusion

I started by considering the general agentive assumption according to which the body that one experiences as one's own is the body represented in the hot body map. This account, however, fails to explain certain findings. I then provided a finer-grained account of the notion of hot body map and proposed distinguishing the working body map involved in instrumental movements from the protective body map involved in self-defence. What the protective body map highlights is the fact that the body has a unique significance for the organism's evolutionary needs. The bodyguard hypothesis thus solves the difficulties that a more general agentive conception faces, while doing justice to the intuition that the sense of bodily ownership is intimately related to action. The function of the protective body map is indeed both to draw the boundary of the territory to defend and to guide how to best defend it, and what better defence than to directly connect what is to be protected with the means for protecting it? The protective body map represents both the body that is to be guarded and the bodyguard itself.

10

The Narcissistic Body

This final chapter develops the bodyguard hypothesis in more detail by bringing all the pieces together to paint a consistent—if not exhaustive—picture of the sense of bodily ownership. In particular, I argued in Chapter 1 that there is a specific feeling in virtue of which one experiences one's body as one's own. But if one grants that there is such a feeling, the question then becomes: what is it like to feel one's body as one's own? Now that we know the grounds of the sense of bodily ownership, we are in a situation to answer this question.

Let me start with a first approximation by drawing a parallel with *nationalistic feelings*. Consider what was called the Russian "patriotic war" of 1812. In the nineteenth century, people who lived in the Russian Empire were characterized by the heterogeneity of their cultures and religions and the enormous distances that separated them. It was only at the time when Napoleon invaded Russia and burnt Moscow down that the Russians felt that they belonged to the same country. More generally, it has been noted in social psychology that feeling one's country as one's own involves not just being aware of its borders, nor being aware that one can vote; it involves being aware that what happens to the country directly matters, and this feeling is especially salient when the country is under threat. One can then experience a feeling of national unity against the common enemy.

As I shall now argue, to some extent this characterization applies to the sense of bodily ownership: experiencing one's body as one's own does not simply involve being aware of its boundaries, nor does it merely involve being aware that one can control it; it involves being aware of its special significance thanks to the protective body map that represents it. Thanks to the protective body map too, one can experience one's body as being unified rather than as a patchwork of bodily pieces: a threat to a specific body part is a threat to the whole body and to protect it requires the whole body to act together. In this chapter, I shall describe this phenomenology of bodily ownership in more detail. I shall then reply to some objections that could be made against the bodyguard hypothesis.

10.1 What It Is Like to Feel My Body as My Own

In Chapter 9 I focused my attention on the sensorimotor underpinning of the sense of bodily ownership. Let us now turn to its phenomenology: how should one characterize the specific experience that one has when one is aware of one's body as one's own? Martin (1992, p. 201) said: "What marks out a felt limb as one's own is not some special quality that it has, but simply that one feels it in this way". I will now reply to him: there is a special quality that marks out a felt limb as one's own, namely an affective quality that arises from the protective frame of reference of bodily experiences. After ruling out several possible interpretations of the notion of affective quality, I will defend the view that it is best understood in terms of narcissistic feeling.

10.1.1 What Affective Quality?

One way to interpret the bodyguard hypothesis is to assume that the sense of bodily ownership consists in *bodily care*. Along with other forms of attachments, such as love, care can be defined as a kind of sentiment (e.g. Broad 1954; Frijda 2007; Deonna and Teroni, 2012). Sentiments consist in long-term affective dispositions that structure the subject's multifaceted relationship with their focus.

On the cognitive side, ascriptions of sentiments go hand in hand with dispositions to make distinctive evaluations of the situations involving the object of the sentiment. A lover will for instance apprehend situations involving her beloved in a way not shared by those lacking her sentiment [. . .] On the motivational side, we conceive of someone with a given sentiment as having a specific motivational structure: the lover is motivated to further the beloved's interests, to help him when in need, to act in the beloved's interests even when they conflict with hers, etc. [. . .] Sentiments are also dispositions to feel specific types of emotions in given actual or counterfactual circumstances.   (Deonna and Teroni, 2012, p. 109)

More specifically, the sentiment of care gives import to objects and warrants the emotions that one can experience about them as well as specific behaviours (Helm, 2002): if I care about my Ming vase, I will pay special attention to what surrounds it, be ready to catch it if it falls, and be angry at you if you throw a baseball at it, and so forth. Or if I care about my body, I will be ready to withdraw my hand from the hot stove when I feel pain (Bain, 2014; Klein, 2015b):

> Pains motivate because we care about our bodies. Were we to stop caring, something that's nearly impossible, for good biological reasons, then pains wouldn't matter.   (Klein, 2015b, p. 500)

The description of bodily care thus seems to account for a number of features described in Chapter 9, including the broad consistent pattern of behavioural, affective, and attentional responses associated with the protective body map. If this is the right interpretation of the bodyguard hypothesis, then the body that one feels as one's own is the body that one cares about.

The problem with this interpretation is that the sentiment of caring is too broad to account for the special relationship that one normally has only with one's own body. One cares about many things, and sometimes even more than one cares about one's own body. For many indeed, there is little doubt that their children matter more than themselves. The notion of bodily care may also seem too cognitively rich and complex. As I described it, the protective body map is a nonconceptual representation that arises from a low-level mechanism that is biologically rooted. It may contribute to the sentiment of bodily care that we experience but it seems likely that many other factors contribute to this sentiment.

I suggest instead that one should appeal to the thinner notion of *affective feelings* in order to shed light on the phenomenology of ownership. To explore this specific affective phenomenology, I will take as a starting point one "familiar" affective feeling, namely the feeling of familiarity. I see my students entering the classroom. The phenomenology of my experience includes the visual phenomenology of the colour of their eyes and of the shape of their face, but it includes something more. When I see them at the end of the academic term, I am aware that I know them: they look familiar. The feeling of familiarity can be defined as a specific type of affective phenomenology elicited by the perception of objects and events that have personal significance. It cannot be reduced to the recognition of the visual features of the face but involves autonomic responses, which result in increased arousal (Bauer, 1984; Tranel and Damasio, 1985; Ellis and Lewis, 2001). The phenomenology of my visual experience is thus dual, both sensory and affective (Dokic and Martin, 2015). Because of this duality, it is possible to have preserved sensory phenomenology with no affective phenomenology. This is what happens in Capgras syndrome. Patients with this syndrome can see that a person is visually identical to their spouse for example, but they do not feel that she is their spouse and they believe that this person must be an impostor. Their sensory phenomenology is thus intact, but they lack the affective responses normally associated with it, as shown by the absence of arousal when they look at familiar faces: visual recognition is preserved, but not autonomic recognition (Ellis et al., 1997). Their delusion of an impostor is only an attempt to explain their "incomplete" perceptual experiences of their spouse. By contrast, patients with Frégoli delusion have an anomalously heightened affective responsiveness for unknown individuals, and thus believe that they are surrounded by familiar persons in disguise (Langdon et al., 2014).

Interestingly, the feeling of familiarity calls to mind certain features of the sense of bodily ownership. Broadly speaking, they both concern the perception of bodies that have a special significance. Another interesting point is that the SCR is used both as a measure of familiarity in face recognition, and as a measure of ownership in most RHI studies. Finally, there is an apparent similarity between somatoparaphrenia and Capgras syndrome: patients suffering from those disorders experience a body (their own or someone else's) as unfamiliar and novel. A somatoparaphrenic patient, for instance, complained: "My hand is not like this, this is different, it's too short." (Gandola et al., 2012, p. 1176). Feinberg and Roane (2003) thus conclude that somatoparaphrenia should be viewed as a kind of Capgras syndrome for one's body parts.

However, despite these similarities, the sense of bodily ownership cannot be reduced to the feeling of familiarity. Firstly, the function of the feeling of familiarity is not to track exclusively our own body: our body feels familiar, but so too do many bodies. One might still try to spell out the phenomenology of ownership in terms of a specific kind of familiarity feeling, what may be conceived of as a feeling of "extreme" familiarity. However, this feeling would still lack the specific motivational force that characterizes the sense of bodily ownership, which has a clear positive valence: it motivates you to protect the body that has such significance. By contrast, the feeling of familiarity has no positive or negative valence: both your friend and the enemy that you fight can feel all too familiar. Hence, the feeling of familiarity does not fully capture the phenomenology of bodily ownership. Nonetheless the feeling of familiarity illustrates the fact that the significance of objects and events for the subject can give rise to specific affective feelings. The significance that familiarity expresses is simply not of the same type as the significance that ownership expresses. The former results from previous encounters with the person, whereas the latter results from selective pressure. I thus suggest interpreting the latter in narcissistic terms.

10.1.2 The Narcissistic Quality of Bodily Experiences

By qualifying bodily experiences as being *narcissistic*, I do not mean that they involve a kind of self-love that eventually ends badly.[1] As just argued, the affective phenomenology that characterizes the sense of ownership is thinner than any emotion or sentiment. Instead, I use the notion of narcissism partly in the same way as Akins (1996) does in her analysis of the function of sensory systems.

Akins's hypothesis is that sensory systems are not just servile detection systems that aim to be as reliable as possible in carrying information about the states with which they co-vary. Rather, sensory systems have what she calls a narcissistic function: they aim at securing what is best for the organism. Narcissistic perception is not about what is perceived but about the impact of what is perceived for the subject. This theory of perception can already be found in Descartes's *Sixth Meditation*: "These perceptions of the senses, although given me by nature, merely to signify to my mind what things are beneficial and hurtful to the composite whole of which it is a part". A similar view can also be found in Malebranche (1674). The senses do not represent "immutable truths that preserve the life of the mind" but "mutable things proper to the preservation of the body" (Malebranche, 1674 p. 261).[2] To illustrate her

---

[1] Narcissus fell in love with his reflection without being aware that it was himself that he loved but later realized this: "I am he. I sense it and I am not deceived by my own image. I am burning with love for myself." (Ovid, Bk III: 437–73). But his love for himself eventually killed him.

[2] For more detail on Descartes's and on Malebranche's views, see Simmons (2008).

hypothesis, Akins appeals to the case of thermal sensations, which indicate what is safe or dangerous for the body given its thermal needs:

What the organism is worried about, in the best of narcissistic traditions, is its own comfort. The system is not asking, "What is it like out there?", a question about the objective temperature states of the body's skin. Rather it is doing something—informing the brain about the presence of any relevant thermal events. Relevant, of course, to itself.   (Akins, 1996, p. 349)

Here I do not adopt Akins's whole theoretical framework. In particular, I do not defend the view that all sensory systems are narcissistic. Rather, I claim that more than any other type of perceptual experiences, bodily experiences—including thermal sensations and possibly even more interoceptive feelings—obey narcissistic principles. By being narcissistic, bodily awareness is not about the body *simpliciter*; it is about the body for the self. The protective body map clearly fulfils a narcissistic function by informing the brain about the potential relevance of the location of the sensation for the organism's needs. Hence, if a spider crawls on my hand, I feel its contact as being located within the frame of the body to protect. Thanks to their protective reference frame, bodily experiences involve the awareness of the body as having a special import for the self. They are thus endowed with a specific affective colouring that goes beyond their sensory phenomenology. The narcissistic quality of bodily experiences cannot be reduced to the sensory recognition of bodily properties, but involves autonomic responses. This narcissistic quality constitutes the phenomenology of bodily ownership.

There are two ways such an affective colouring of perceptual experiences might be understood in representational terms. Firstly, it can be conceived of as an affective experience distinct from the sensory experience that it is bound to (Dokic and Martin, 2015). Alternatively, it can be conceived of as a specific affective mode of presentation of the sensory content of perceptual experiences (Matthen, 2005). To settle the debate, one needs to ask whether one can experience the narcissistic quality independently of any bodily experiences. In other words, can the protective body map directly give rise to the feeling of ownership independently of the bodily experiences that it spatially structures?

Consider the following study by Paqueron et al. (2003) who analysed what subjects reported after anaesthesia. They found that half of the subjects spontaneously reported that their anaesthetized limb felt "dead" but only five out of the thirty-six participants reported that it felt as if the anaesthetized limb did not belong to them. A further study of patients with spinal cord injury shows that some but not all patients experience a loss of ownership of their deafferented limbs (Scandola et al., 2017). One way to interpret these results is to say that the absence of bodily sensations does not always preclude the feeling of bodily ownership, which must thus be independent. The difficulty, however, is that one can never be sure of the complete absence of bodily sensations. All somatosensory inputs can be blocked and yet one can experience phantom sensations. We can further note that if the feeling of bodily presence is a prerequisite for ownership, as argued in Chapter 2, then there will never be narcissistic feelings independently of any bodily experiences.

One encounters similar difficulties with the analysis of other affective feelings. For instance, Dokic and Martin (2015) suggest that the feeling of déjà vu should be interpreted as a free-floating feeling of familiarity without any associated sensory experience. If this is the correct interpretation, then feelings of familiarity can be fully independent from sensory experiences. However, Dokic's and Martin's claim seems too strong: the feeling of déjà vu is not completely disconnected from any sensory experience; it is bound to the global sensory content of a scene, although one does not know which particular object or event in this scene triggers the feeling. I thus doubt that one can empirically settle the debate between the two interpretations of affective feelings. It then becomes purely a conceptual issue on the nature of phenomenology, which goes beyond the scope of this book.

10.1.3 A Matter of Degrees?

We can now reinterpret cases of somatoparaphrenia in which patients can still feel bodily sensations located in the body part that feels as alien. As in Capgras syndrome, these patients have their sensory phenomenology preserved, while their affective one is missing. On the other hand, patients with embodiment delusion also have their sensory phenomenology preserved, while their affective one is misguided (directed towards the wrong body), and should thus be compared to Fregoli delusion. These cases, however, are extreme ones, and one may wonder whether in between there might be degrees of bodily ownership in the same way as there are degrees of familiarity. As William James (1890,

p. 242) notes, "certain parts of the body seem more intimately ours than the rest". The RHI, for instance, would be an intermediary case, in which the rubber hand feels to be less our own than our biological hand, but still more our own than another individual's hand. One might also suggest that depending on the degrees of disownership that one experiences, one is either delusional or not.

However, at the phenomenological level, what distinguishes patients that are delusional from those that are not is primarily their *feeling of confidence*. One of the defining features of delusions is indeed the unshakable certainty in the delusional belief so that it is maintained against all evidence. A doctor once asked his patient suffering from somatoparaphrenia, "From zero to ten, how much are you sure this is not your hand?". The patient replied, "Ten." (Invernizzi et al., 2013, p. 149). Hence, one can be more or less confident that this is not one's hand. Likewise, ownership ratings in questionnaires reflect variation in the feeling of confidence experienced by the subjects in their sense of ownership: participants in the RHI experiments show only weak confidence, whereas we normally show high confidence. Differences in degrees of confidence, however, do not entail differences in degrees of ownership itself.

Still one may argue that the significance for survival of the different parts of one's body is not a matter of all or nothing. Roughly speaking, the little finger seems to be less worthy of protection than the index finger. If the sense of bodily ownership expresses this special significance, then it should also be continuous rather than discrete. Degrees of significance would then correspond to degrees of ownership. This analysis, however, is not at the right level. It is not as if to each ounce of flesh were ascribed a value for survival. As argued, the protective body map conceives of the body at a more holistic level. It represents structural affordances, which are integrative units of effectors brought together by the movements that they allow the organism to perform. It does not matter if this bit of the body is less essential than that other bit as long as they are all integrated in the body in action.

Let us finally rule out a last possible reason for which one might be tempted to conceive of bodily ownership in terms of degrees. Arguably, the protective body map results from the Bayesian pondering of cues and factors of various types, which computes the posterior probability, that is, the degree of belief in the prior hypothesis conditioned on the observation of sensory evidence.[3] The question now is whether it *feels* the same when the posterior probability is 0.9 (there is little doubt that this is my hand), and when it is 0.75 (it is likely that this is my hand)? Here I believe that to put too much emphasis on probability in computational models can be misleading. First, it is not clear that these probabilities have any psychological reality. According to many, Bayesian frameworks only aim at characterizing the task that the mind has to perform but they do not make a priori commitments about the actual algorithmic computations in the brain (e.g. Griffith et al., 2010). Secondly, even if indeed the system computes probabilities, it does not stop there. Computational models are supposed to apply to sensory systems in general but the content of most of our perceptual experiences is not in terms of probabilities. For example, I visually experience a line as being vertical; I do not experience it as having a probability of 0.9 of being vertical. This is so because the sensory system must make a *decision* on the basis of posterior probability in order to act or to make a perceptual judgement. The decision requires further computations using gain and loss functions. For every decision, there are specific consequences, which can be positive or negative. The rule aims at maximizing the expected gain of the decision given the posterior probability. The outcome of the decision is then discrete. What I am suggesting is that the experience of ownership should be analysed at the level of the binary decision (is this my hand or not?), rather than at the level of the graded posterior probability.

---

[3] A Bayesian model starts with some a priori knowledge about how a system should work given biological and environmental constraints. It is represented by a prior probability distribution for a model's structure and parameters—what the variables are and how they influence each other. It aims at computing the posterior probability, which is proportional to the product of the prior probability and the likelihood function. The likelihood is the probability of the data given the hypothesis. It represents everything that one knows about the process that turns the state of the world into sensory information. In the case that concerns us, once the posterior probability is computed, the system has an estimate of whether a body part is represented in the protective body map. The probability for it to be part of the protective body map is thus the result of the interplay between the inputs, prior knowledge that models the body and the world, and expectation for uncertainty due to internal noise in the sensorimotor system and external noise in the world. When the uncertainty is expected to be high, then the weight of inputs is downgraded in favour of prior knowledge about the body. I am extremely grateful to Pascal Mamassian for his clarifying expertise on computational models of perception. See Mamassian et al. (2002), for instance.

I experience this hand as mine; I do not experience it as having a probability of 0.9 that it is mine.

To recapitulate, the phenomenology of ownership should be conceived of in affective terms. It expresses the awareness of the narcissistic significance of the body. Such awareness normally results from the frame of reference of bodily experiences that is based on the protective body map. In the absence of this protective frame, one can still have bodily experiences but one no longer experiences the body that one feels as being one's own.

10.2 The Body in Danger

From now on I will consider various objections that a critic might be tempted to make against the bodyguard hypothesis. I will start with the most intuitive one. Clearly, one does not always protect the body that one experiences as one's own and one protects many things—including many bodies—besides one's own body without having a sense of ownership for them. Hence, the body that one experiences as one's own cannot be the body that one protects. I agree with the conclusion, but as I will now show, it does not invalidate the bodyguard hypothesis.

10.2.1 Judgement of Muscles and Judgements of Reasons

It is important at this point to recapitulate what the bodyguard hypothesis exactly claims. It simply assumes that the body map that grounds the sense of bodily ownership plays a motivational role for protective behaviours; it does not assume that it is the only factor that decides which body is to be protected. Like any other behaviour, protective behaviours can result from complex decision-making processes, involving a variety of beliefs, desires, emotions, moral considerations, and so forth. There are thus many situations in which we do not protect our body and yet still experience it as our own. It can be for altruistic reasons, for instance. In a famous study on empathy, participants had to watch on a screen a student receiving ten electric shocks, but after two shocks they saw the student showing extreme distress and the experimenter telling her that she could stop if they were willing to trade places with the student (Batson et al., 1981). Participants with high empathy score agreed to do this and thus to protect the student and put themselves at risk. However, their prosocial behaviour did not entail that their protective body map was impaired. It simply shows that other considerations, including empathy, overcame its motivational force.

Furthermore, there are different levels at which protective behaviours can be analysed. Consider the pleasure that some experience in extreme sports. Even if the mountain biker is ready to go downhill on a very steep and dangerous slope, he also pays extreme attention to its immediate environment and he is ready to react in case of obstacles. Or consider the case of the soldier ready to sacrifice himself for his country. He would be a bad soldier of little use to his nation if he were not extremely vigilant and if he let himself be killed at the first attack. In all these cases, the protective body map is preserved even if at some level one puts one's body at risk.

Here it may be useful to distinguish between two types of motivational forces, which are respectively based on what Cherry (1986) calls judgements of reason and judgements of muscle. Judgements of reason can apply to many objects and persons, and it results from the appraisal of various considerations, which can be easily altered depending on the situation. For example, there is no doubt that I protect my laptop because of its professional, personal, and financial value, but I would no longer protect it if these reasons no longer held, if it was broken for instance. According to Cherry, however, there is another kind of judgement that targets exclusively one's own body, which he describes as muscle-based: it is biologically rooted and does not need to be justified by further attitudes. Consequently, it does not depend on the context: there is no situation in which one's body can no longer be needed (I do not normally cease to protect it just because it is "broken", for instance):

Judgments of muscles declare in favour of the self [. . .] [they] are the outward and urgent expression of an ever-accompanying 'I', an ever-present notional personal concern.   (Cherry, 1986, p. 303)

The protective body map gives rise to muscle-based commands but we can sometimes disobey those commands. If one really wants to put the bodyguard to the test, one needs to focus on cases of consistent failure to protect one's body over time, when even judgements of muscle seem to be completely lacking.

10.2.2 Fearless and Painless Bodies

After bilateral amygdala lesion patients experience no fear. The most studied of these patients, SM, was threatened several times in her life, including by a man who wanted to stab her, but she did not react (Tranel and Damasio, 1989). Furthermore, she displays abnormal processing of her peripersonal space, and does not exhibit discomfort when another person is very close to her body (Kennedy et al., 2009). Patients who suffer from pain asymbolia also do not react to threat or to pain. They seem to be in pain insofar as they are able to judge the location and the intensity of painful stimuli, but they do not experience pain as unpleasant nor do they try to avoid it. And when they are injured, they show none of the standard reactions, like screaming or withdrawing their hand from the harmful stimulus (Grahek, 2001). These two types of patients seem to show no inclination to protect their own bodies, and yet, as far as I know, they still experience their body as their own. Do they invalidate the bodyguard hypothesis?

What is interesting is that their affective attitude towards danger is not simply neutral. Actually, it has been found that SM shows increased arousal under threat (Tranel and Damasio, 1989). This might indicate that at some basic level her body still matters to her or that she is simply interested by what is happening. But in both interpretations, she is not indifferent. And she does not merely fail to run away from danger, she is actually eager to face it. For instance, although she said that she hated snakes, when exposed to them she was visually captivated by them and held a snake for over three minutes while rubbing it and touching its tongue. She repeatedly commented, "This is so cool!" (Feinstein et al., 2011). Likewise, patients with pain asymbolia expose themselves to noxious stimuli with a smile:

On occasion, the patient willingly offered his hands for pain testing and laughed during stimulation. He had no concern about the defect and appeared highly cooperative during pain evaluation.   (Berthier et al., 1988, pp. 42–3)

The fact that these patients have no fear and seem to enjoy painful stimulations to some extent or to look for danger suggests that they suffer from something different from a deficit of protective body map. Instead, one may suggest that they are *misevaluating* what is going on. Ramachandran (1998, p. 1858) proposes the following "story" for pain asymbolia:

One part of the person's brain (the insula) tells him, "here is something painful, a potential threat" while another part (the cingulate gyrus of the limbic system) says a fraction of a second later, "oh, don't worry, this is no threat after all."

A similar story can be told about patients with amygdala lesions. The amygdala plays a crucial role in detection and prioritization of threat-related information (LeDoux, 2000; Bach et al., 2014), and its impairment in SM can cause her to receive contradictory information about the badness of what is going on (what is dangerous is evaluated as interesting, for instance). Therefore, she does not protect her body from it but rather explores it. The failure to engage in protective behaviour that is observed in these syndromes might result from the fact that individuals are deficient in the detection of threats rather than from impairments to their protective body map.

10.2.3 Too Much Pain?

We have just considered the case of patients who do not care if they are in pain. We shall now consider the case of patients who are in extreme pain and who do care. These patients are interesting for our purpose because they often no longer refer to the body part in pain by using the first-person pronoun and they can report a sense of disownership (Morse and Mitcham, 1998). For instance, a prisoner reported: "It's like the torture starts and as it gets stronger I just get further away from the body" (Ataria, 2015, p. 205). Patients who suffer from complex regional pain syndrome (CRPS) also often describe a sense of disconnection from the affected limb: "It was just like this foreign body you were carrying around with you cause it didn't feel like it was part of you" (Lewis et al., 2007, p. 114).[4] By detaching themselves from their bodies, all these patients hope to detach themselves from their pains. As the neurologist Schilder (1935 p. 104) noted: "When the whole body is filled with pain, we try to get rid of the whole body".

Arguably, the body part in pain is still represented in the protective body map in these patients (it may actually be over-represented, Moseley, 2005): they show no self-harming behaviour and they can be extremely vigilant in protecting it from further pain. Yet they report a sense of disownership. On the basis of such cases, one might conclude

---

[4] Complex regional pain syndrome is a chronic pain condition that often happens after an injury like a broken arm or nerve lesions, and that causes intense burning pain associated with a decreased ability to move the affected body part, with swelling and stiffness in affected joints. CRPS can sometimes be conceived of as the mirror phenomenon of pain asymbolia: chronic pain is mainly driven by the affective component of pain (Bliss et al. 2016), which is said to be lacking in pain asymbolia.

that the protective body map does not suffice for experiencing one's body as one's own. This conclusion, however, is too hasty because one should be cautious in the interpretation of their disownership reports. On a scale for 0 to 10, with 10 corresponding to "It feels like I own it completely", patients with CRPS actually still rate their sense of ownership of the affected hand on average at 5.9 (Moseley et al., 2012b). It is less than what healthy participants report, but it is still above average. As mentioned earlier, one should interpret such ratings in terms of confidence, which appear to be relatively weak here. But this is different from actually denying ownership of the body part in pain. Furthermore, this rating may simply reflect not what they experience but what they desire. A patient thus commented: "I didn't *want* to make it feel like mine in a lot of ways because it hurt so much" (Lewis et al., 2007, p. 116, my emphasis). At some level, the patients thus disavow a part of their body when in intense pain but this does not entail that they no longer experience it as their own.

10.2.4 Self-Destructive Behaviours

Let us finally consider the tragic cases of individuals who voluntarily injure themselves or commit suicide. The objective here is not to provide a comprehensive understanding of self-destructive behaviours. As said earlier, many factors contribute to protective behaviours and to their absence. The objective is only to assess whether the bodyguard hypothesis is compatible with the will to destroy one's body. It might seem indeed that in many suicidal individuals there is little left of their protective body map. For instance, their reaction to pain is different, showing stronger pain resistance and appraising the pain as less intense than control groups (e.g. Adler and Gataz, 1993; Dworkin, 1994; Orbach et al., 1996):

In the suicidal group, the mounting inner stress results in an increase in the thresholds for external stimulation, probably as a defensive detachment, resulting in greater exposure to further harm and eventually self-destructive behavior. This kind of defensive adjustment by suicidal individuals may facilitate self-destruction, as it increases the exposure to internal and external danger.

(Orbach et al., 1996, p. 317)

In other words, their body does not seem to matter to them any more. According to the bodyguard hypothesis, the sense of bodily ownership in suicidal individuals should also be affected. But is it the case?

Unfortunately, little attention has been given to the way they experience their body. Nonetheless, it has been suggested that dissociative mechanisms intervene in suicidal behaviour as a kind of defence strategy, which results in detachment from the body (Orbach et al., 1996; Russ et al., 1993). Furthermore, we know that some suicidal individuals also suffer from schizophrenia, a psychiatric condition in which bodily awareness is disturbed (e.g. Klaver and Dijkerman, 2016). Many other suicidal individuals suffer from major depression, in which depersonalization is conceived of as one of the five main signs. In particular, they describe that their body feels like an empty shell external to them and that it is unable to obey them, unable to move. They thus want to get rid of it. Here are some reports from several patients:

"pNR3: I feel my body as heavy, depressed, inert. It endures gravity. It is a nuisance."

"pR6: I have the sensation that I have a heavy body that moves with difficulty and a mind that is external to the body and that becomes agitated and panics."

"pR7: My dream would be to be able to dispense with the body. My body, that's what is too much, that's what hinders." (Gouzien, 2016, p. 56–7, my translation)

In its most extreme form, depression can lead to the Cotard syndrome, in which patients describe their body as being literally dead, as a rotting corpse. In all these cases, the protective body map is most probably altered, but so is the sense of bodily ownership.

One should finally note that self-injuries might also be a way for some patients to regain their sense of ownership. In Chapter 9 we saw that patients with congenital pain insensitivity could experience their body as an extraneous tool. Interestingly, these patients often engage in self-mutilation, including burns and auto-amputations of fingertips and tongues (Danziger and Willer, 2009; Nagasako et al., 2003). Similar self-inflicted injuries can also be found in animals raised in isolation preserved from pain (Melzack and Scott, 1957; Lichstein and Sackett, 1971). Injuring one's body may be a way to test whose body it is: feeling something can then strengthen a fading feeling of confidence in the sense of bodily ownership, whereas feeling nothing can confirm that this body has nothing to do with one. Or it might be that one simply feels no sense of ownership whatsoever: one is not voluntarily injuring one's own body; one is simply injuring *a body*, which happens to be one's own. In short, the disruption of the protective body map in turn can result in self-inflicted injuries (Borah et al., 2016).

10.3 A Threat of Circularity

So far I have failed to find any valid empirical counterexample to the bodyguard hypothesis. But one might still criticize it at the conceptual level and argue that it cannot account for the first-personal character of the sense of bodily ownership. To recall, I do not refer here to the subjectivity of the phenomenology of ownership (i.e. *I* feel this body as my own). I refer to the fact that the phenomenology of bodily ownership normally grounds self-ascriptive judgements (e.g. this body is *my own*). What is at the origin of this first-personal character? On the face of things, the bodyguard hypothesis seems to confront a dilemma that was pointed out by Peacocke (2015) against my earlier account of ownership in terms of body schema (Vignemont, 2007). If the protective body map represents one's body *qua* one's own, then it presupposes what it is supposed to explain, but if it does not, then one is left with no explanation of the first-personal character of the sense of bodily ownership. I shall now consider each horn of the dilemma.

The narcissistic quality expresses the value of the body for the self but it may seem that this specific body is valuable for the self in virtue of being experienced as one's own. If so, the bodyguard hypothesis faces a version of the classic Euthyphro dilemma. Plato noted that an action is just because it pleases the gods, but the action pleases the gods because it is just. Similarly, it might seem that I experience my body as my own because it has a special significance for me, but for my body to have such significance presupposes that I experience it as mine, or at least that I represent it as mine. Put another way, if the protective body map represents one's body *qua* one's own, then it can account for the first-personal character of the sense of bodily ownership, but it does so "by taking for granted the notion of ownership by a subject, rather than by offering some kind of reductive explanation of the notion" (Peacocke, 2015, p. 174).

However, the bodyguard hypothesis only assumes that the function of the protective body map is to represent the body that matters for the organism's survival; it does not assume that it represents the body that matters for the organism's survival *qua* one's own, even in a nonconceptual way. This distinction is important if one wants to block the risk of circularity. Consequently, being aware that this is my body presupposes that my body has a special affective significance for me, and for my body to have such significance is for it to be the body to protect for the organism's evolutionary needs. Because biology provides an independent standard to ground the notion of significance that is used by the body-guard hypothesis, there is no circularity. As Rey (1977) describes it:

> Thus we may take seriously the kind of concern each person normally exhibits about all and only the person with whom she is identical. That there is such kind of concern I shall not pause here to doubt. It very probably arises *from our biology.* (Rey, 1977, p. 45, my emphasis)

The bodyguard hypothesis can thus avoid the first horn of Peacocke's dilemma, but it may then seem to succumb to the second one: since the protective body schema itself has no first-person component, how can it be the origin of the first-personal character of the sense of bodily ownership? To show that the bodyguard hypothesis has the resources to provide such an account we might want to revisit Akins's notion of narcissism.

According to Akins, the narcissistic question can be phrased as follows: "But how does this all relate to ME?" (Akins, 1996, p. 345). On her view, this question does not only affect the content of my experiences, filtering only what is relevant to me; it also marks the structure, or the format, of my experiences, like a signature: "by asking the narcissistic question, the *form* of the answer is compromised: it always has a self-entered (sic) glow" (Akins, 1996, p. 345). The notion of self-centred glow calls to mind perspectival experiences. For instance, when I see a tree, I experience the location of the tree in its *spatial* relation to me and my egocentric experience "there is a tree in front" can ground self-locating beliefs of the type "I am facing a tree". Likewise, when I feel a spider on my hand, I experience the location of the spider in its *narcissistic* relation to me and my bodily experience "I feel a spider on the body that matters" can ground ownership judgements of the type "I feel a spider on *my* body" because the function of the protective body map is to track my own body.

To recapitulate, it is a fact of the matter that there is a specific body that one should protect to survive and reproduce and it is the function of the protective body map to reliably covariate with it. The protective body map is normally recruited as a spatial frame of reference for bodily experiences, ascribing a narcissistic value to the body that one experiences. One is then aware of one's body as one's own and one is motivated to protect it (Figure 10.1). In some rare cases, however, the protective

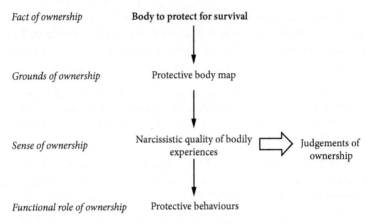

Fact of ownership   **Body to protect for survival**

Grounds of ownership   Protective body map

Sense of ownership   Narcissistic quality of bodily ⟹ Judgements of
                                    experiences   ownership

Functional role of ownership   Protective behaviours

Figure 10.1. The bodyguard hypothesis.

body map malfunctions. Then it can either incorporate extraneous limbs or objects, leading to pathological or illusory sense of ownership and heightened affective response, or exclude a part of one's biological body, leading to pathological sense of disownership and diminished affective response.

10.4 Conclusion

The bodyguard hypothesis has succeeded in explaining a number of empirical findings and allowed us to account for a number of features of bodily awareness. In particular, we can now reply to the following questions:

- *What grounds the sense of bodily ownership?* The sense of bodily ownership is normally grounded in the spatial frame of reference of bodily experience that is provided by a specific type of hot body map, namely the protective body map.
- *What does it mean to experience one's body as one's own?* Feeling one's body as one's own is to experience a specific type of affective phenomenology that can be described in narcissistic terms.
- *What is the function of the sense of bodily ownership?* The sense of bodily ownership motivates and guides self-protection. It highlights at the conscious level that this body is to be guarded. Roughly speaking, it expresses: "Don't touch! Beware! Don't get too close!".

Appendix 1

Bodily Illusions

Given the recent booming of empirical research on bodily awareness, it is impossible to provide an exhaustive review of all the findings. Here I will simply describe some of the main bodily illusions that I refer to in the book.

A.1 The Rubber Hand Illusion and Some of Its Many Versions

A.1.1 The Rubber Hand Illusion (RHI)

In the classic set-up, one sits with one's arm hidden behind a screen, while fixating on a rubber hand presented in one's bodily alignment; the rubber hand is then stroked in either synchrony or asynchrony with one's hand (Botvinick and Cohen, 1998). The illusion, which occurs only in the synchronous condition, includes the following components (see Table A.1):

Table A.1. The rubber hand illusion

Phenomenological level *(measured by questionnaires)*	Referred sensations	Participants report that they feel tactile sensations as being located, not on their real hand that is stroked but on the rubber hand.
	Sense of ownership	They report feeling as if the rubber hand belonged to them, were part of their body, and were their hand.
Behavioural level	Proprioceptive drift	They mislocalize the finger that was touched in the direction of the location of the finger of the rubber hand.
Physiological level	Arousal	When they see the rubber hand threatened, they display an increased affective response (as measured by their skin conductance response).

A.1.2 The Fake Finger Illusion

The first version of the RHI can actually be found as early as 1937 in the work of the French psychologist Tastevin (1937). A fake finger is aligned with one's hand, which is hidden behind a cloth. Then one closes one's eyes, while the fake finger is moved farther away. When one opens one's eyes, one cannot see one's real finger, but one can see the fake finger and one feels as if one's own finger were located where the fake finger is, although the two fingers are up to 30 cm apart.

A.1.3 The Mirror Illusion

When one views one's static left arm in the mirror as if it were one's right arm, but at a location further away than the real location of one's right arm, one mislocalizes one's right hand towards the reflection (Holmes et al., 2004).

A.1.4 The Tool Ownership Illusion

One sits with one's arm hidden behind a screen, while fixating on a mechanical grabber presented in one's bodily alignment; the "fingers" of the grabber are then stroked in either synchrony or asynchrony with the fingers of one's hand. After synchronous stimulation, one feels tactile sensations to be located on the tool and one reports that one feels as if the tool were part of one's body. Furthermore, one mislocalizes one's hand in the direction of the location of the grabber and one shows an increase in one's physiological response when the grabber is threatened significantly more after synchronous than asynchronous stroking. The ownership report and the affective response are not increased after using the tool (Cardinali et al., under revision).

A.1.5 The Enfacement Illusion

The experimenter strokes both one's face and the face of another individual who is seated in front. One is then shown the morphed picture of the two faces and asked to judge whose face it is. After synchronous stimulation, one judges more often that it is one's own face than after asynchronous stimulation (Tsakiris, 2008; Sforza et al., 2010).

A.1.6 The Invisible Hand Illusion

One sits with one's arm hidden behind a screen, while fixating on the table. The experimenter then synchronously or asynchronously strokes the hidden hand and a discrete volume of empty space 5 cm above the table in direct view. One then localizes one's sensation of touch at the empty location and one reports that it seemed as if one had an "invisible hand" (Guterstam et al., 2013).

A.1.7 The Supernumerary Hand Illusion

Whereas in the classic version of the RHI one's own hand remains hidden from sight, in the supernumerary hand illusion, one can see both one's hand and the rubber hand being stroked. One then reports feeling sensations in the two hands. Furthermore, when asked whether it seems as if one had two right hands, one agrees (Guterstam et al., 2011).

A.1.8 The Somatic Version of the RHI

While blindfolded, one is stroked on one's right hand while stroking a rubber hand in synchrony with one's left hand. One then reports feeling that one is touching one's own right hand. When asked to point to one's right hand after the illusion, one mislocalizes it in the direction of the location of the rubber hand (Ehrsson et al., 2005). This illusion does not work with blind participants (Petkova et al., 2012).

A.1.9 The Moving RHI

While blindfolded, one moves the index finger up and down hidden from sight while one sees the index of the rubber hand performing the same movement either in synchrony or in asynchrony. When asked to point to one's hand after the illusion, one mislocalizes it in the direction of the location of the rubber finger and one reports feeling the rubber finger moving (e.g. Kalckert and Ehrsson, 2012; Tsakiris et al., 2006).

A.2 Full-Body Illusions and Some of Its Many Versions

These bodily illusions use the same principles of visual capture of touch as the RHI but manipulate the experience of the *whole body*.

A.2.1 The Full-Body Illusion

With a system of video cameras and virtual reality goggles one has the illusory experience of seeing one's *back*, as if one were behind oneself. One then sees one's back being stroked while feeling it being stroked. When the strokes are synchronous, one reports feeling the stroke on the body seen in the distance. In addition, one reports that one feels as if the seen body were one's own. When one is passively moved and asked to walk back to one's original location, one walks too far in the direction of the location of the seen body (Lenggenhager et al., 2007).

A.2.2 The Out-of-Body Illusion

With a system of video cameras and virtual reality goggles one has the illusory experience of seeing one's *chest*, as if one were in front of oneself. One then sees the chest being stroked, while feeling it being stroked. When the strokes are synchronous, one reports feeling that one is located where the seen chest is and that one is touched on it. One also has a strong affective reaction when the seen body is threatened (Ehrsson, 2007).

A.2.3 The Body-Swapping Illusion

This is another version of the out-of-body illusion. With a double system of video cameras and virtual reality goggles (one for the participant and one for the experimenter), one experiences the illusion of having the experimenter's visual perspective so that one sees her body from a first-person perspective, as if it were one's own. One is then asked to stand opposite the experimenter, to take hold of her hand, and to squeeze it. One can thus see the experimenter's hand and one's biological hand shaking, but one's visual perspective is from the experimenter's point of view. After synchronous squeezing, a participant reported: "I was shaking hands with myself!" (Petkova and Ehrsson, 2008, p. 5). Interestingly, the physiological response is stronger when the experimenter's hand is threatened than when it is one's own biological hand that is under threat.

A.3 Other Bodily Illusions

A.3.1 The Numbness Illusion

One holds one's palm against another person's palm and strokes with the index and thumb of the contralateral hand the two joined index fingers. One then reports different alterations of body perception, such as a sensation of numbness and widening of one's own finger, as well as a feeling of owning the other person's index finger (Dieguez et al., 2009).

A.3.2 The Disownership Illusion

One places one's right hand inside a box that allows seeing live video images of one's hand. One is then asked to reach across to touch one's right hand with one's left hand. The experimenter then creates a situation in which the right hand unexpectedly disappears from vision and touch: one's right hand has slowly moved outwards without one noticing it. All that can be seen and felt is the table where the right hand had once been. One then reports feeling that the right hand is no longer part of one's body and one does not react when one sees the experimenter pretending to stab the last seen location of one's right hand or its real hidden actual location (Newport and Gilpin, 2011; see also Gentile et al., 2013).

A.3.3 *The Japanese Illusion*

One crosses one's wrists, one's hands clasped with thumbs down. Then one turns one's hands in towards one's chest until one's fingers point upward. Then someone touches one of one's fingers. One then has difficulty in moving the finger that was touched, but also in reporting which finger it was.

A.3.4 *The Thermal Grill Illusion*

Originally discovered by Thunberg in 1896, this illusion induces a sensation of painful heat by touching interlaced harmless levels of warm and cool bars to the skin (for instance, index finger warm, middle finger cold, ring finger warm).

A.3.5 *The Cutaneous Rabbit Illusion*

Repeated rapid tactile stimulation to the wrist, then near the elbow, can create the illusion of touches at intervening locations along the arm, as if a rabbit were hopping along it.

A.3.6 *The Parchment Skin Illusion*

One reports feeling that one's skin is dry, parchment-like, after hearing a distorted recording of the sounds produced by one's hands rubbing back and forth (with high frequencies accentuated) (Jousmaki and Hari, 1998).

A.3.7 *The Marble-Hand Illusion*

One reports feeling one's hand as stiffer, heavier, and harder after hearing the sound of a hammer against the skin progressively replaced by the sound of a hammer hitting a piece of marble (Senna et al., 2014).

A.3.8 *The Dentist Illusion*

After dental surgery, one's mouth often feels bigger, despite the fact that it looks normal. It has been shown that when the sensory input from the lips and front teeth is fully blocked following anaesthesia, one can feel one's lips and teeth increasing in size by as much as 100 per cent (Gandevia and Phegan, 1999; Türker et al., 2005).

A.3.9 *The Pinocchio Illusion*

If the tendons of one's arm muscles are vibrated at a certain frequency, one experiences illusory arm movements. One feels, for instance, one's arm moving away if one's biceps tendon is vibrated, and if one simultaneously grasps one's nose, one experiences one's nose as elongating by as much as 30 cm. The Pinocchio illusion results from a sensorimotor conflict between erroneous proprioceptive information (i.e. one's arm moving away) and accurate tactile information (i.e. contact between one's nose and one's fingers) (Lackner, 1988).

Appendix 2

Neurological and Psychiatric Bodily Disorders

Bodily disorders are not rare. There are very few pure cases but they often accompany psychiatric and neurological syndromes (after brain lesion, peripheral lesion, migraine, and epileptic seizure) to such an extent that it has been found that half of the stroke patients in a neurological department was impaired at least at one level of bodily awareness (Denes, 1990; Schwoebel and Coslett, 2005). There seems to be no dimension of bodily awareness that cannot be disrupted. Here is a description of the main disorders.

Alice in Wonderland syndrome Distorted awareness of the size, mass, shape of the body, or its position in space (including macro/microsomatognosia and OBE).

Allesthesia See Allochiria.

Allochiria Mislocalization of sensory stimuli (tactile, visual, auditory) to the corresponding opposite half of the body or space.

Allodynia Pain due to a stimulus that does not normally produce pain.

Alien hand sign See Somatoparaphrenia.

Anarchic hand sign Unintended but purposeful and autonomous movements of the upper limb and intermanual conflict.

Anorexia nervosa Eating disorder characterized by self-starvation.

Anosodiaphora Lack of concern for one's deficits.

Anosognosia Lack of awareness of one's deficits, such as hemiplegia.

Apotemnophilia See Xenomelia.

Asomatognosia See Somatoparaphrenia.

Autoscopy Experience of seeing one's body in extrapersonal space.

Autoprosopagnosia Inability to recognize one's own face.

Autotopagnosia Mislocalization of body parts and bodily sensations.

Body integrity identity disorder (BIID) See Xenomelia.

Body-specific aphasia Loss of lexical knowledge of body parts.

Bulimia nervosa Eating disorder characterized by recurrent binge eating, followed by compensatory behaviour.

Complex regional pain syndrome (CRPS) Chronic pain condition usually after an injury or trauma to a limb characterized by prolonged or excessive pain and mild or dramatic changes in skin colour, temperature, and/or swelling in the affected area.

Congenital pain insensitivity Inability to experience pain from birth.

Conversion disorder (hysteria) Functional disorder with no organic cause.

Cotard syndrome Delusional belief that one is dead, does not exist, is putrefying, or has lost one's blood or internal organs.

Deafferentation Loss of tactile and proprioceptive information.

Delusional parasitosis (or Ekbom syndrome) Tactile hallucination associated with the delusion that small creatures are infesting one's skin.

Depersonalization Altered, detached, or estranged subjective experience.

Dyschiria See Allochiria.

Dysmorphophobia Distorted perception of one's self-appearance.

Embodiment delusion Feeling another individual's left arm as being one's own.

Fading limb Lack of awareness of the presence and position of the limb if not seen.

Finger agnosia Inability to individuate and recognize the fingers.

Gerstmann's syndrome Finger agnosia, agraphia, acalculia, and left-right confusion.

Heautoscopy Seeing a double of oneself at a distance.

Hemi-depersonalization Feeling of absence or of separation of body parts.

Heterotopagnosia Designation of parts of the body of another person when asked to point towards one's own body.

Hyperalgesia Increased response to a stimulus that is normally painful.

Hypochondrias Excessive somatic concern.

Ideomotor apraxia Inability to execute or carry out skilled movements and gestures.

Internal heautoscopy Visual experience of one's inner organs.

Interoceptive agnosia Loss of pain feeling.

Macro/Micro-somatognosia Distorted awareness of the size of the whole body or of body parts (bigger or smaller).

Medically unexplained symptoms Somatoform illness.

Mirror sign Inability to recognize one's own image in the mirror.

Misoplegia Hatred towards one's own body parts.

Motor neglect Underutilization of one side of the body.

Munchausen syndrome Chronic factitious disorder, in which a person repeatedly acts as if she had a bodily disorder.

Negative heautoscopy Inability to see one's reflection in a reflecting surface.

Numbsense Loss of tactile awareness with preserved tactually guided movements.

Out of the body experience (OBE) Visual awareness of one's own body from a location outside the physical body.

Pain asymbolia Lack of affective and motor responses to nociceptive sensations.

Personal neglect Lack of attention towards one side of the body.

Phantom limb Awareness of a non-existing limb, often after amputation.

Pseudopolymyelia Delusion of reduplication of body parts.

Pusher syndrome Postural deviation towards the contralesional side.

Somatoparaphrenia Denial of ownership of one's own body part.

Supernumerary limb Illusory experience of a supplementary limb.

Synchiria Visual or tactile stimulation on the ipsilesional side of the body resulting in the perception of stimuli on both the ipsilesional and contralesional side.

Tactile extinction Lack of awareness of tactile stimuli on the contralesional limb during simultaneous bilateral stimulation.

Vestibular disorder Dysfunction of the vestibular system that can induce vertigo, loss of balance, and blurred vision.

Xenomelia Urge to be amputated of one's own perfectly healthy limb.

Appendix 3

Somatoparaphrenia

Somatoparaphrenia generally occurs after a lesion in the right parietal cortex. It is also sometimes called *asomatognosia* (Feinberg, 2009). It can also be referred to as the "main étrangère" or *the alien hand sign* (Brion and Jedynak, 1972; Marchetti and Della Salla, 1998) but it should not be confused with the anarchic hand. The *anarchic hand* refers to the experience of autonomous semi-purposeful movements of the arm that are experienced as independent of the patient's volition. The *alien hand* refers to a feeling of estrangement between the patient and one of her hands when the patient touches her hand without visual control. One of the first descriptions was given by Jean-Baptiste Bouillaud (1825) about a patient who said about the left side of his body that it felt "as if it were a stranger to him; it seemed to him that somebody else's body was lying on his side, or even a corpse" (p. 64, translated by Bartolomeo et al., 2017). Later Hermann Zingerle (1913) reported the case of a patient who persistently stated "that to the left of himself a woman was lying in the bed" (Benke et al., 2004, p. 268). Since then many reports have been given of patients who deny ownership of one of their limbs as their own. The "alien" limb can then be attributed to another individual (to the examiner, to relatives, and even to dead persons). Alternatively, they personify it: it has "a mind of its own" (Leiguarda et al., 1993), it is a "buddy" (Feinberg et al., 1998), or a "monster" (Vallar and Ronchi, 2009). Finally, they may sometimes attribute to themselves another person's hand (i.e. embodiment delusion). Given the diversity among the cases, it has been suggested to use the generic label of "disturbed sensation of a limb ownership" (Baier and Karnath, 2008) to refer to all of them but this term is heavily theoretically loaded. It assumes that (i) there are sensations of bodily ownership and (ii) that they are missing in these disorders. However, as discussed in Chapter 1, both assumptions are controversial. Here I use the term of somatoparaphrenia for the sake of neutrality and simplicity.

Somatoparaphrenia is often—but not always—associated with motor and somatosensory deficits, spatial neglect, anosognosia for hemiplegia, and the anarchic hand sign. However, it cannot be reduced to any of these disorders (Moro et al., 2004; Vallar and Ronchi, 2009). It may be more frequent than is usually assumed because it can disappear very quickly. In one study, 61% of patients with hyperacute right middle cerebral artery infarct display somatoparaphrenia at onset, evolving to 15% after a week (Antoniello and Gottesman, 2017).

Here is a sample of introspective reports from somatoparaphrenic patients.

Feeling of disownership	*"How am I supposed to know whose hand is this? It's not mine."* (Gandola et al., 2012, p. 1176) *"My eyes and my feelings don't agree, and I must believe my feelings. I know they look like mine, but I can feel they are not, and I can't believe my eyes."* (Nielsen, 1938, p. 555)
Feeling of confidence	*"E: From zero to ten, how much are you sure this is not your hand? P: Ten."* (Invernizzi et al., 2013, p. 149) *"I don't know. Maybe it is mine. But no, I'm sure, it isn't mine, I don't feel it as my hand."* (Cogliano et al., 2012, p. 764)
Attribution to others	*"It's my niece's hand." "She is so kind, she left it here to keep me company (. . .) she's also very absent-minded. Look here, she was rushing home and forgot her hand here."* (Gandola et al., 2012, p. 1177) *"E: Whose hand is this? P: Yours! E: Mine? Are you sure? P: Yes, sure, whose hand is it supposed to be?"* (Invernizzi et al., 2013, p. 149).
Feeling of presence	*"It occupies a lot of space in this bed, it's so uncomfortable (. . .) it is very intrusive, I've no more space in this bed!"* (Gandola et al., 2012, p. 1176) *"E: The nurses told us that you woke up and called them. Why? P: because there was this hand here."* (Invernizzi et al., 2012, p. 148)
Feeling of familiarity	*"My hand is not like this, this is different, it's too short."* (Gandola et al., 2012, p. 1176) *"E: Isn't it a little bit weird to have a foreign hand with you? P: No! My hand is not like this! This is shorter, plus it does nothing!!"* (Invernizzi et al., 2012, p. 148)
Agentive feeling	*"This arm, it does not move, it does not obey. It doesn't want to do anything, it doesn't help me. It betrays me."* (Cogliano et al., 2012, p. 766) *"If you want it, I'll give it to you as my gift, since I have no need for it. It doesn't work. Maybe you'll be able to get it working."* (Gandola et al., 2012, p. 1177)
Bodily care	*"ML: I said: 'I got to get rid of them'. TEF: Yes; ML: So I did; TEF: So what did you do? ML: Put them in a garbage."* (Feinberg, 2009, p. 15) *"Yes, please take it away. I don't care about its destiny as it is not mine."* (Gandola et al., 2012, p. 1176) *"Because it does as it likes. It is sick. I would very much like to throw it away."* (Cogliano et al., 2012, p. 766)

References

Abdulkarim, Z. and Ehrsson, H. H. (2016). No causal link between changes in hand position sense and feeling of limb ownership in the rubber hand illusion. *Attention, Perception, and Psychophysics*, 78(2), 707–20.

Adler, G. and Gataz, W. F. (1993). Pain perception threshold in major depression. *Biological Psychiatry*, 34, 687–9.

Akins, K. (1996). Of sensory systems and the "aboutness" of mental states. *The Journal of Philosophy*, 93(7), 337–72.

Alary, F., Duquette, M., Goldstein, R., Elaine Chapman, C., Voss, P., La Buissonnière-Ariza, V., and Lepore, F. (2009). Tactile acuity in the blind: a closer look reveals superiority over the sighted in some but not all cutaneous tasks. *Neuropsychologia*, 47(10), 2037–43.

Alsmith, A. J. T. (2012). The concept of a structural affordance. *Avant: The Journal of the Philosophical-Interdisciplinary Vanguard*, 3, 2, 94–107.

Alsmith, A. J. T. (2015). Mental activity and the sense of ownership. *Review of Philosophy and Psychology*, 6, 4, 881–96.

Andersen, E. S. (1978). Lexical universals of body-part terminology. In J. H. Greenberg (ed.), *Universals of human language*. Stanford, CA: Stanford, CA University Press, pp. 335–68.

Anderson, M. L. (2007). The massive redeployment hypothesis and the functional topography of the brain. *Philosophical Psychology*, 20.2: 143–74.

Anderson, M. L. (2010). Neural reuse: a fundamental organizational principle of the brain. *Behavioral and Brain Sciences*, 33, 4, 245–313.

Anema, H. A., van Zandvoort, M. J., de Haan, E. H., Kappelle, L. J., de Kort, P. L., Jansen, B. P. et al. (2009). A double dissociation between somatosensory processing for perception and action. *Neuropsychologia*, 47, 1615–20.

Anscombe, G. E. M. (1957). *Intention*. Ithaca: Cornell University Press.

Anscombe, G. E. M. (1962). On sensations of position. *Analysis*, 22(3), 55–8.

Anscombe, G. E. M. (1965). The intentionality of sensation: A grammatical feature. In Ronald J. Butler (ed.), *Analytic Philosophy*. Oxford: Blackwell, pp. 158–80.

Antoniello, D. and Gottesman, R. (2017). Limb misidentification: a clinical-anatomical prospective study. *The Journal of Neuropsychiatry and Clinical Neurosciences*, 29(3): 284–8.

Apps, M. A. and Tsakiris, M. (2013). The free-energy self: a predictive coding account of self-recognition. *Neuroscience and Biobehavioral Reviews*, 41: 85–97.

Armel, K. C. and Ramachandran V. S. (2003). Projecting sensations to external objects: evidence from skin conductance response. *Proceedings of the Royal Society Biology Sciences*, 270(1523), 1499–506.

Armstrong, D. (1962). *Bodily sensations*. London: Routledge and Paul.

Arntzen, C. and Elstad, I. (2013). The bodily experience of apraxia in everyday activities: a phenomenological study. *Disability and Rehabilitation*, 35(1), 63–72.

Artaud, A. (1948). Pour en finir avec le jugement de Dieu. Paris: Gallimard.

Asai, T. (2015). Illusory body-ownership entails automatic compensative movement: for the unified representation between body and action. *Experimental Brain Research*, 233(3), 777–85.

Aspell, J. E., Lenggenhager, B., and Blanke, O. (2009). Keeping in touch with one's self: multisensory mechanisms of self-consciousness. *PloS One*. 4(8):e6488.

Ataria, Y. (2015). Sense of ownership and sense of agency during trauma. *Phenomenology and the Cognitive Sciences*, 14: 199–212.

Auclair, L., Noulhiane, M., Raibaut, P., and Amarenco, G. (2009). Where are your body parts? A pure case of heterotopagnosia following left parietal stroke. *Neurocase*, 15(6), 459–65.

Auvray, M. and Spence, C. (2008). The multisensory perception of flavor. *Consciousness and Cognition*, 17(3), 1016–31.

Avenanti, A., Annela, L., and Serino, A. (2012). Suppression of premotor cortex disrupts motor coding of peripersonal space. *Neuroimage*, 63(1), 281–8.

Avenanti, A., Bueti, D., Galati, G., and Aglioti, S. M. (2005). Transcranial magnetic stimulation highlights the sensorimotor side of empathy for pain. *Nature Neuroscience*, 8(7), 955–60.

Avenanti, A., Minio-Paluello, I., Sforza, A., and Aglioti, S. (2009). Freezing or escaping? Opposite modulations of empathic reactivity to the pain of others. *Cortex* 45(9): 1072–7.

Aymerich-Franch, L. and Ganesh, G. (2016). The role of functionality in the body model for self-attribution. *Neuroscience Research*, 104, 31–7.

Baas, U., de Haan, B., Grässli, T., Karnath, H. O., Mueri, R., Perrig, W. J., Wurtz, P., and Gutbrod, K. (2011). Personal neglect—A disorder of body representation? *Neuropsychologia*, 49(5), 898–905.

Bach, D. R., Guitart-Masip, M., Packard, P. A., Miró, J., Falip, M., Fuentemilla, L., and Dolan, R. J. (2014). Human hippocampus arbitrates approach-avoidance conflict. *Current Biology*, 24(5), 541–7.

Bach, P., Peatfield, N. A., and Tipper, S. P. (2007). Focusing on body sites: the role of spatial attention in action perception. *Experimental Brain Research* 178(4): 509–17.

Badde, S. and Heed, T. (2016). Towards explaining spatial touch perception: weighted integration of multiple location codes. *Cognitive Neuropsychology* 33(1–2): 26–47.

Baier, B. and Karnath, H. O. (2008). Tight link between our sense of limb ownership and self-awareness of actions. *Stroke*, 39(2): 486–8.

Bain, D. (2003). Intentionalism and pain. *Philosophical Quarterly*, 53(213), 502–23.

Bain, D. (2013). What makes pains unpleasant? *Philosophical Studies*, 166: S69–S89.

Bain, D. (2014). Pains that Don't Hurt. *Australasian Journal of Philosophy* 92(2): 1–16.

Banissy, M. J. and Ward, J. (2007). Mirror-touch synaesthesia is linked with empathy. *Nature Neuroscience* 10: 815–81.

Bartolomeo, P., de Vito, S., and Malkinson, T. S. (2017). Space-related confabulations after right hemisphere damage. *Cortex*, 87: 166–73.

Bassolino, M., Finisguerra, A., Canzoneri, E., Serino, A., and Pozzo, T. (2015). Dissociating effect of upper limb non-use and overuse on space and body representations. *Neuropsychologia*, 70, 385–92.

Bassolino, M., Serino, A., Ubaldi, S., and Làdavas, E. (2010). Everyday use of the computer mouse extends peripersonal space representation. *Neuropsychologia*, 48(3): 803–11.

Batson, C. D., Duncan, B. D., Ackerman, P., Buckley, T., and Birch, K. (1981). Is empathic emotion a source of altruistic motivation? *Journal of Personality and Social Psychology*, 40(2), 290.

Bauer, R. M. (1984). Autonomic recognition of names and faces in prosopagnosia: A neuropsychological application of the guilty knowledge test. *Neuropsychologia*, 22(4), 457–69.

Bayne, T. (2008). The phenomenology of agency. *Philosophy Compass*, 3(1), 182–202.

Bayne, T. (2010). Agentive experiences as pushmi-pullyu representations. In J. Aguilar, A. Buckareff, and K. Frankish (eds), *New waves in the philosophy of action*. London: Palgrave Macmillan, pp. 219–36.

Bayne, T. and Levy, N. (2005). Amputees by choice: body integrity identity disorder and the ethics of amputation. *Journal of Applied Philosophy*, 22(1), 75–86.

Bayne, T. and Pacherie, E. (2007). Narrators and comparators: the architecture of agentive self-awareness. *Synthese*, 159: 475–91.

Beck, B. B. (1980). *Animal Tool Behavior: The use and manufacture of tools*. New York: Garland Press.

Bellan, V., Gilpin, H. R., Stanton, T. R., Newport, R., Gallace, A., and Moseley, G. L. (2015). Untangling visual and proprioceptive contributions to hand localisation over time. *Experimental Brain Research*, 233(6): 1689–701.

Benke, T., Luzzatti, C., and Vallar, G. (2004). Hermann Zingerle's "Impaired perception of the own body due to organic brain disorders": an introductory comment, and an abridged translation. *Cortex*, 40(2), 265–74.

Bennett, D. and Hill, C. (eds) (2014). *Sensory integration and the unity of consciousness*. Cambridge, MA: MIT Press.

Berger, M. and Gerstenbrand, F. (1981). Phantom illusions in spinal cord lesion. In J. Siegfried and M. Zimmermann (eds), *Phantom and stump pain*. Berlin: Springer, pp. 66–73.

Berlucchi, G. and Aglioti, S. M. (1997). The body in the brain: neural bases of corporeal awareness. *Trends in Neuroscience*, 20(12), 560–4.

Berlucchi, G. and Aglioti, S. M. (2010). The body in the brain revisited. *Experimental Brain Research*, 200(1), 25–35.

Bermúdez, J. L. (1998). *The Paradox of Self-Consciousness*. Cambridge, MA: MIT Press.

Bermúdez, J. L. (2005). The phenomenology of bodily awareness. In D. Woodruff Smith and A. Thomasson (eds), *Phenomenology and Philosophy of Mind*. Oxford: Clarendon Press, 295–316.

Bermúdez, J. L. (2011). Bodily awareness and self-consciousness. In S. Gallagher (ed.), *Oxford Handbook of the Self*. Oxford: Oxford University Press.

Bermúdez, J. L. (2015). Bodily ownership, bodily awareness, and knowledge without observation. *Analysis*, 75(1): 37–45.

Bermúdez, J. L. (2017). Ownership and the space of the body. In F. de Vignemont and A. Alsmith (eds), *The subject's matter: self-consciousness and the body*. Cambridge, MA: MIT Press.

Bernardi, N. F., Marino, B. F., Maravita, A., Castelnuovo, G., Tebano, R., and Bricolo, E. (2013). Grasping in Wonderland: altering the visual size of the body recalibrates the body schema. *Experimental Brain Research*, 226(4): 585–94.

Berthier, M., Starkstein, S., and Leiguarda, R. (1988). Asymbolia for pain: a sensory-limbic disconnection syndrome. *Annals of Neurology*, 24/1: 41–9.

Berti, A. and Frassinetti, F. (2000). When far becomes near: remapping of space by tool use. *Journal of Cognitive Neuroscience*, 12(3), 415–20.

Biederman, I. (1987). Recognition-by-components: a theory of human image understanding. *Psychological Review*, 94, 115–47.

Billon, A. (2017). Mineness first. In F. de Vignemont and A. Alsmith (eds), The Subject's Matter: self-consciousness and the body. Cambridge, MA: MIT Press.

Bisiach, E. and Berti, A. (1995). Consciousness in dyschiria. In M. Gazzniga (ed.), *The cognitive neurosciences*. Cambridge, MA: MIT Press.

Bisiach, E., Rusconi, M. L., and Vallar, G. (1991). Remission of somatoparaphrenic delusion through vestibular stimulation. *Neuropsychologia*, 29(10): 1029–31.

Bisiach, E., Vallar, G., Perani, D., Papagno, C., and Berti, A. (1986). Unawareness of disease following lesions of the right hemisphere: anosognosia for hemiplegia and anosognosia for hemianopia. *Neuropsychologia*, 24(4): 471–82.

Blakemore, S.-J., Bristow, D., Bird, G., Frith, C., and Ward, J. (2005). Somato-sensory activations during the observation of touch and a case of vision-touch synaesthesia. *Brain*, 128, 1571–83.

Blanke, O., Slater, M., and Serino, A. (2015). Behavioral, neural, and computational principles of bodily self-consciousness. *Neuron*, 88(1): 145–66.

Blankenburg, F., Ruben, J., Meyer, R., Schwiemann, J., and Villringer, A. (2003). Evidence for a rostral-to-caudal somatotopic organization in human primary somatosensory cortex with mirror-reversal in areas 3b and 1. *Cerebral Cortex*, 13(9): 987–93.

Bliss, T. V., Collingridge, G. L., Kaang, B. K., and Zhuo, M. (2016). Synaptic plasticity in the anterior cingulate cortex in acute and chronic pain. *Nature Reviews Neuroscience*, 17(8): 485–96.

Block, N. (1983). Mental pictures and cognitive science. *Philosophical Review*, 92, 499–541.

Block, N. (2005). Review of Alva Noë, Action in perception, *The Journal of Philosophy*, CII: 259–72.

Bolognini, N., Ronchi, R., Casati, C., Fortis, P., and Vallar, G. (2014). Multisensory remission of somatoparaphrenic delusion: my hand is back! *Neurology: Clinical Practice*, 4, 216–25.

Bonnier, P. (1905). L'Aschématie [Aschematia]. *Revue Neurologique* (Paris), 13, 605–9.

Borah, S., McConnell, B., Hughes, R., and Kluger, B. (2016). Potential relationship of self-injurious behavior to right temporo-parietal lesions. *Neurocase*, 22(3): 269–72.

Bottini, G., Bisiach, E., Sterzi, R., and Vallar, G. (2002). Feeling touches in someone else's hand, *Neuroreport*, 13, 249–52.

Botvinick, M. (2004). Neuroscience. Probing the neural basis of body ownership. *Science*, 305(5685): 782–3.

Botvinick, M. and Cohen, J. (1998). Rubber hands 'feel' touch that eyes see. *Nature*, 391, 756.

Bouillaud, J. B. (1825). *Traité clinique et physiologique de l'encéphalite, ou inflammation du cerveau, et de ses suites*. Paris: J.-B. Baillière.

Braam, A. W., Visser, S., Cath, D. C., and Hoogendijk, W. J. (2006). Investigation of the syndrome of apotemnophilia and course of a cognitive-behavioural therapy. *Psychopathology*, 39(1): 32–7.

Bradshaw, M. (2016). *A return to self: depersonalization and how to overcome it*, Seattle, WA: Amazon Services International.

Brass, M., Bekkering, H., and Prinz, W. (2001). Movement observation affects movement execution in a simple response task. *Acta Psychologica* (Amst), 106(1–2): 3–22.

Brass, M. and Heyes, C. (2005). Imitation: Is cognitive neuroscience solving the correspondence problem? *Trends in Cognitive Sciences*, 9, 489–95.

Brecht, M. (2017). The body model theory of somatosensory cortex. *Neuron*, 94(5): 985–92.

Bremner, A. J. (2017). The origins of body representations in early life. In F. de Vignemont and A. Alsmith, *The subject's matter: self-consciousness and the body*. Cambridge, MA: MIT Press.

Bremner, A. J., Mareschal, D., Lloyd-Fox, S., and Spence, C. (2008). Spatial localization of touch in the first year of life: early influence of a visual spatial code and the development of remapping across changes in limb position. *Journal of Experimental Psychology General*, 1371: 149–62.

Bremner, A. J. and Spence, C. (2017). The development of tactile perception. *Advances in Child Development and Behavior*, 52: 227–68.

Brewer, B. (1995). Bodily awareness and the self. In J. L. Bermúdez, T. Marcel, and N. Eilan (eds), *The body and the self*. Cambridge, MA: MIT Press.

Brion, S. and Jedynak, C. P. (1972). Troubles du transfert interhémisphérique callosal disconnection. A propos de trois observations de tumeurs du corps calleux. Le signe de la main étrangère. *Revue Neurologique*, 126, 257–66.

Briscoe, R. E. (2009). Egocentric patial representation in action and perception. *Philosophy and Phenomenological Research* 79(2): 423–60.

Broad, C. D. (1954). Emotion and sentiment. In D. Cheeney (ed.), *Broad's critical essays in moral philosophy*. New York: George Allen & Unwin.

Brooks, R. A. (1991). Intelligence without representation. *Artificial Intelligence*, 47(1–3): 139–59.

Brown, C. H. (1976). General principles of human anatomical partonomy and speculations on the growth of partonomic nomenclature. *American Ethnologist*, 3, 400–24.

Brown, C. H. (2005). Hand and arm. In M. Haspelmath, M. Dryer, D. Gil, and B. Comrie (eds), *The world atlas of language structures*. Oxford: Oxford University Press, pp. 522–5.

Brozzoli, C., Makin, T. R., Cardinali, L., Holmes, N. P., and Farnè, A. (2012). Peripersonal space: a multisensory interface for body–object interactions. In M. M. Murray and M. T. Wallace (eds). *The Neural Bases of Multisensory Processes*. Boca Raton, FL: CRC Press.

Brugger, P., Kollias, S. S., Müri, R. M., Crelier, G., Hepp-Reymond, M.-C., and Regard, M. (2000). Beyond re-membering: phantom sensations of congenitally absent limbs. *Proceedings of the National Academy of Sciences of the United States of America*, 97(11): 6167–72.

Buccino, G., Binkofski, F., Fink, G. R., Fadiga, L., Fogassi, L., Gallese, V., Seitz, R. J., Zilles, K., Rizzolatti, G., and Freund, H.-J. (2001). Action observation activates premotor and parietal areas in a somatotopic manner: An fMRI study. *European Journal of Neuroscience*, 13, 400–4.

Bufalari, I., Aprile, T., Avenanti, A., Di Russo, F., and Aglioti, S. (2007). Empathy for pain and touch in the human somatosensory cortex. *Cerebral Cortex*, 17(11): 2553–61.

Burin, D., Livelli, A., Garbarini, F., Fossataro, C., Folegatti, A., Gindri, P., and Pia, L. (2015). Are movements necessary for the sense of body ownership? Evidence from the Rubber Hand Illusion in pure hemiplegic patients. *PLoS One*, 10(3): e0117155.

Butler, S. (1872). *Erewhon*. London: Penguin Classics.

Buxbaum, L. J., Giovannetti, T., and Libon, D. (2000). The role of the dynamic body schema in praxis: evidence from primary progressive apraxia. *Brain and Cognition*, 44(2): 166–91.

Cameron, B. D., de la Malla, C., and López-Moliner, J. (2015). Why do movements drift in the dark? Passive versus active mechanisms of error accumulation. *Journal of Neurophysiology*, 114(1): 390–9.

Campbell, J. (1999). Schizophrenia, the space of reasons, and thinking as a motor process. *The Monist*, 82: 4, 609–25.

Campbell, J. (2002). *Reference and consciousness*. Oxford: Clarendon Press.

Campbell, J. (2006). What is the role of location in the sense of a visual demonstrative? Reply to Matthen. *Philosophical Studies*, 127, 239–54.

Cantello, R., Civardi, C., Cavalli, A., Varrasi, C., and Vicentini, R. (2000). Effects of a photic input on the human cortico-motoneuron connection. *Clinical Neurophysiology* 111: 1981–9.

Canzoneri, E., Magosso, E., and Serino, A. (2012). Dynamic sounds capture the boundaries of peripersonal space representation in humans. *PLoS One*, 2012; 7(9): e44306.

Canzoneri, E., Marzolla, M., Amoresano, A., Verni, G., and Serino, A. (2013a). Amputation and prosthesis implantation shape body and peripersonal space representations. *Scientific Reports,* 3: 2844.

Canzoneri, E., Ubaldi, S., Rastelli, V., Finisguerra, A., Bassolino, M., and Serino, A. (2013b). Tool-use reshapes the boundaries of body and peripersonal space representations. *Experimental Brain Research*, 228(1), 25–42.

Cardinali, L., Brozzoli, C., and Farnè, A. (2009b). Peripersonal space and body schema: two labels for the same concept? *Brain Topography*, 21(3–4): 252–60.

Cardinali, L. Brozzoli, C., Luauté, J., Roy, A. C., and Farnè, A. (2016). Proprioception is necessary for body schema plasticity: evidence from a deafferented patient. *Frontiers in human neuroscience*, 10.

Cardinali, L., Brozzoli, C., Urquizar, C., Salemme, R., Roy, A. C., and Farnè, A. (2011). When action is not enough: tool-use reveals tactile-dependent access to Body Schema. *Neuropsychologia*, 49(13): 3750–7.

Cardinali, L., Frassinetti, F., Brozzoli, C., Urquizar, C., Roy, A. C., and Farnè, A. (2009a). Tool-use induces morphological updating of the body schema. *Current Biology*, 19(12), R478–R479.

Cardinali, L., Roy, A. C., Yanofsky, R., de Vignemont, F., Culham, J. C., and Farnè, A. (under revision). Tool-being: mapping one's hand based on tool vision and function.

Carroll, L. (1865). *Alice in Wonderland*. London: Macmillan & Co.

Cassam, Q. (1995). Introspection and bodily self-ascription. In J. L. Bermúdez, T. Marcel, and N. Eilan (eds), *The body and the self*. Cambridge, MA: MIT Press.

Cassam, Q. (1997). *Self and world*. New York: Oxford University Press.

Chao, L. L. and Martin, A. (2000). Representation of manipulable man-made objects in the dorsal stream. *Neuroimage*, 12(4), 478–84.

Chemero, T. (2009). *Radical embodied cognitive science*. Cambridge, MA: MIT Press.

Cherry, C. (1986). Mine and mattering. *Philosophy and Phenomenological Research*, 47, 2, 297–304.

Clark, A. (1989). *Microcognition: philosophy, cognitive science, and parallel distributed processing*. Cambridge, MA: MIT Press.

Clark, A. (2001). Visual experience and motor action: Are the bonds too tight? *Philosophical Review*, 110(4), 495–519.

Clark, A. (2004). Feature-placing and proto-objects. *Philosophical psychology*, 17(4): 443–69.

Clark, A. (2008). *Supersizing the mind: embodiment, action, and cognitive extension*. Oxford: Oxford University Press.

Cleret de Langavant, L., Trinkler, I., Cesaro, P., and Bachoud-Lévi, A. C. (2009). Heterotopagnosia: when I point at parts of your body. *Neuropsychologia*, 47(7), 1745–55.

Cléry, J., Guipponi, O., Wardak, C., and Hamed, S. B. (2015). Neuronal bases of peripersonal and extrapersonal spaces, their plasticity and their dynamics: knowns and unknowns. *Neuropsychologia*, 70: 313–26.

Coburn, R. C. (1966). Pain and space. *Journal of Philosophy*, 63(13), 381–96.

Coello, Y., Bartolo, A., Amiri, B., Devanne, H., Houdayer, E., and Derambure, P. (2008). Perceiving what is reachable depends on motor representations: evidence from a transcranial magnetic stimulation study. *PLoS One*, 3(8): e2862.

Cogliano, R., Crisci, C., Conson, M., Grossi, D., and Trojano, L. (2012). Chronic somatoparaphrenia: a follow-up study on two clinical cases. *Cortex*, 48(6): 758–67.

Cole, J. (1995). *Pride and a daily marathon*. Cambridge, MA: MIT Press.

Coltheart, M., Langdon, R., and McKay, R. (2011). Delusional belief. *Annual Review of Psychology*, 62: 271–98.

Cook, R., Press, C., Dickinson, A., and Heyes, C. (2010). Acquisition of automatic imitation is sensitive to sensorimotor contingency. *Journal of Experimental Psychology: Human Perception and Performance*, 36(4): 840–52.

Coslett, H. B. (1998). Evidence for a disturbance of the body schema in neglect. *Brain and Cognition*, 37, 527–44.

Costantini, M., Ambrosini, E., Tieri, G., Sinigaglia, C., and Committeri, G. (2010). Where does an object trigger an action? An investigation about affordances in space. *Experimental Brain Research*, 207(1–2), 95–103.

Craig, A. D. (2003). Interoception: the sense of the physiological condition of the body, *Current Opinion in Neurobiology*, 13: 500–5.

Cronholm, B. (1951). Phantom limbs in amputees: a study of changes in the integration of centripetal impulses with special reference to referred sensations. *Acta Psychiatrica Et Neurologica Scandinavica*, 72: 1–310.

Crucianelli, L., Metcalf, N. K., Fotopoulou, A. K., and Jenkinson, P. M. (2013). Bodily pleasure matters: velocity of touch modulates body ownership during the rubber hand illusion. *Frontiers in Psychology*, 4: 703.

Csibra, G. (2007). Action mirroring and action interpretation: an alternative account. In P. Haggard, Y. Rosetti, and M. Kawato (eds), *Sensorimotor foundations of higher cognition. Attention and performance XXII*. Oxford: Oxford University Press.

Curt, A., Yengue, C. N., Hilti, L. M., and Brugger, P. (2011). Supernumerary phantom limbs in spinal cord injury. *Spinal Cord*, 49(5): 588–95.

Cussins, A. (2012). Environmental representation of the body. *Review of Philosophy and Psychology*, 3(1), 15–32.

Damasio, A. (1999). *The feeling of what happens*. London: William Heinemann.

Danziger, N. and Willer, J. C. (2009). Congenital insensitivity to pain. *Revue Neurologique* (Paris), 165(2): 129–36.

Davies, M., Coltheart, M., Langdon, R., and Breen, N. (2001). Monothematic delusions: towards a two-factor account. *Philosophy, Psychiatry, and Psychology*, 8(2), 133–58.

D'Imperio, D., Tomelleri, G., Moretto, G., and Moro, V. (2017). Modulation of somatoparaphrenia following left-hemisphere damage. *Neurocase*.

de Bruin, L. and Gallagher, S. (2012). Embodied simulation, an unproductive explanation: comment on Gallese and Sinigaglia. *Trends in Cognitive Sciences*, 16(2), 98–9.

de Preester, H. and Knockaert, V. (eds). (2005). *Body image and body schema: interdisciplinary perspectives on the body* (Vol. 62). Amsterdam: John Benjamins Publishing.

de Preester, H. and Tsakiris, M. (2009). Body-extension versus body incorporation: is there a need for a body-model? *Phenomenology and Cognitive Sciences*, 8: 307–19.

Dehaene, S. and Cohen, L. (2007). Cultural recycling of cortical maps. *Neuron*, 56: 384–98.

della Gatta, F., Garbarini, F., Puglisi, G., Leonetti, A., Berti, A., and Borroni, P. (2016). Decreased motor cortex excitability mirrors own hand disembodiment during the rubber hand illusion. *eLife*, 5, e14972.

Delk, J. L. and Fillenbaum, S. (1965). Differences in perceived color as a function of characteristic color. *The American Journal of Psychology*, 78(2), 290–3.

Denes, G. (1990). Disorders of body awareness and body knowledge. In F. Boller and J. Grafman (eds), *Handbook of neuropsychology* (Vol. 2). Amsterdam: Elsevier, pp. 207–28.

Deonna, J. A. and Teroni, F. (2012). *The emotions: a philosophical introduction*. Oxford: Routledge.

Descartes, R. (1641/1979). *Les méditations métaphysiques*. Paris: Garnier Flammarion.

Dewe, H., Watson, D. G., and Braithwaite, J. J. (2016). Uncomfortably numb: new evidence for suppressed emotional reactivity in response to body-threats in those predisposed to sub-clinical dissociative experiences. *Cognitive Neuropsychiatry*, 28: 1–25.

Di Pellegrino, G., Fadiga, L., Fogassi, L., Gallese, V., and Rizzolatti, G. (1992). Understanding motor events: a neurophysiological study. *Experimental Brain Research*, 91(1), 176–80.

Di Pellegrino, G., Làdavas, E., and Farnè, A. (1997). Seeing where your hands are. *Nature*, 388, 730.

Di Vita, A., Boccia, M., Palermo, L., and Guariglia, C. (2016). To move or not to move, that is the question! Body schema and non-action oriented body representations: An fMRI meta-analytic study. *Neuroscience and Biobehavioral Reviews*, 68, 37–46.

Di Vita, A., Palermo, L., Piccardi, L., and Guariglia, C. (2015). Peculiar body representation alterations in hemineglect: a case report. *Neurocase*, 21(6), 697–706.

Dieguez, S. and Annoni, I. M. (2013). Asomatognosia: disorders of the bodily self. *The Behavioral and Cognitive Neurology of Stroke*, 2, 170–92.

Dieguez, S., Mercier, M. R., Newby, N., and Blanke, O. (2009). Feeling numbness for someone else's finger. *Current Biology*, 19(24), R1108–R1109.

Dijkerman, H. C. and de Haan, E. H. (2007). Somatosensory processes subserving perception and action. *The Behavioral and Brain Sciences*, 30, 189–201.

Dokic, J. (2003). The sense of ownership: an analogy between sensation and action. In J. Roessler and N. Eilan (eds), *Agency and self-awareness: issues in philosophy and psychology*. Oxford: Oxford University Press.

Dokic, J. (2010). Perceptual recognition and the feeling of presence. In B. Nanay (ed.), *Perceiving the world*. New York: Oxford University Press.

Dokic, J. (2012). Seeds of self-knowledge: noetic feelings and metacognition. In M. J. Beran, J. L. Brandl, J. Perner, and J. Proust (eds), *Foundations of Metacognition*, 302–21.

Dokic, J. and Martin, J.-R. (2015). 'Looks the same but feels different': a metacognitive approach to cognitive penetrability. In A. Raftopoulos and Zeimbekis, J. (eds), *Cognitive effects on perception: new philosophical perspectives*. Oxford: Oxford University Press.

Dretske, F. (2006). Perception without awareness. In Tamar S. Gendler and John Hawthorne (eds), *Perceptual experience*. Oxford: Oxford University Press, 147–80.

Driver, J. and Halligan, P. W. (1991). Can visual neglect operate in object-centred coordinates? An affirmative case-study. *Cognitive Neuropsychology*, 8(6): 475–96.

Driver, J. and Spence, C. (1998). Attention and the cross-modal construction of space. *Trends in Cognitive Sciences*, 2(7), 254–62.

Dugas, L. and Moutier, F. (1911). *La dépersonnalisation*. Paris: Félix Alcan.

Duncan, J. (1984). Selective attention and the organization of visual information. *Journal of Experimental Psychology: General*, 113, 501–17.

Dworkin, H. R. (1994). Pain insensitivity in schizophrenia: a neglected phenomenon and some implications. *Bulletin of Schizophrenia*, 20, 235–48.

Ebisch, S. J., Perrucci, M. G., Ferretti, A., Del Gratta, C., Romani, G. L., and Gallese, V. (2008). The sense of touch: embodied simulation in a visuotactile mirroring mechanism for observed animate or inanimate touch. *Journal of Cognitive Neuroscience*, 20(9): 1611–23.

Ehrsson, H. H. (2007). The experimental induction of out-of-body experiences. *Science*, 317(5841), 1048.

Ehrsson, H. H. (2009). How many arms make a pair? Perceptual illusion of having an additional limb. *Perception*, 38: 310–12.

Ehrsson, H. H., Holmes, N. P., and Passingham, R. E. (2005). Touching a rubber hand: feeling of body ownership is associated with activity in multisensory brain areas. *Journal of Neuroscience*, 25(45): 10564–73.

Ehrsson, H. H., Rosén, B., Stockselius, A., Ragnö, C., Köhler, P., and Lundborg, G. (2008). Upper limb amputees can be induced to experience a rubber hand as their own. *Brain*, 131(12), 3443–52.

Ehrsson, H. H., Wiech, K., Weiskopf, N., Dolan, R. J., and Passingham, R. E. (2007). Threatening a rubber hand that you feel is yours elicits a cortical anxiety response. *Proceedings of the National Academy of Sciences of the United States of America*, 104(23), 9828–33.

Ellis, H. D. and Lewis, M. B. (2001). Capgras delusion: a window on face recognition. *Trends in Cognitive Sciences*, 5(4), 149–56.

Ellis, H. D., Young, A. W., Quayle, A. H., and De Pauw, K. W. (1997). Reduced autonomic responses to faces in Capgras delusion. *Proceedings of the Royal Society of London. Series B: Biological Sciences*, 264(1384), 1085–92.

Engel, A. K., Fries, P., and Singer, W. (2001). Dynamic predictions: oscillations and synchrony in top-down processing. *Nature Review Neuroscience*, 2(10): 704–16.

Ernst, M. O. (2006). A Bayesian view on multimodal cue integration. In G. Knoblich, I. M. Thornton, M. Grosjean, and M. Shiffrar (eds), *Human body perception from the inside out*. New York: Oxford University Press, pp. 105–31.

Evans, G. (1982). *The varieties of reference*. Oxford: Oxford University Press.

Faivre, N., Salomon, R., and Blanke, O. (2015). Visual consciousness and bodily self-consciousness. *Current Opinion in Neurology*, 28(1), 23–8.

Farnè, A. and Ladavas, E. (2000). Dynamic size-change of hand peripersonal space following tool use. *Neuroreport*, 11(8), 1645–9.

Farnè, A., Roy, A. C., Giraux, P., Dubernard, J. M., and Sirigu, A. (2002). Face or hand, not both: perceptual correlates of reafferentation in a former amputee. *Current Biology*, 6: 12(15): 1342–6.

Farnè, A., Serino, A., and Làdavas, E. (2007). Dynamic size-change of peri-hand space following tool-use: determinants and spatial characteristics revealed through cross-modal extinction. *Cortex*, 43(3), 436–43.

Feinberg, T. E. (2009). *From axons to identity: neurological explorations of the nature of the self*. New York: WW Norton.

Feinberg, T. E., DeLuca, J., Giacino J. T., Roane, D. M., and Solms, M. (2005). Right hemisphere pathology and the self: delusional misidentification and reduplication. In T. E. Feinberg and J. P. Keenan (eds), *The lost self: pathologies of the brain and identity*. New York: Oxford University Press.

Feinberg, T. E. and Roane, D. (2003). Misidentification syndromes. In T. E. Feinberg and M. Farah (eds), *Behavioral neurology and neuropsychology* New York: McGraw-Hill, pp. 373–81.

Feinberg, T. E., Roane, D. M., and Cohen, J. (1998). Partial status epilepticus associated with asomatognosia and alien hand-like behaviours. *Archive of Neurology*, 55, 1574–7.

Feinstein, J. S., Adolphs, R., Damasio, A., and Tranel, D. (2011). The human amygdala and the induction and experience of fear. *Current Biology*, 21(1), 34–8.

Felician, O., Ceccaldi, M., Didic, M., Thinus-Blanc, C., and Poncet, M. (2003). Pointing to body parts: a double dissociation study. *Neuropsychologia*, 41(10): 1307–16.

Ferrari, P. F., Rozzi, S., and Fogassi, L. (2005). Mirror neurons responding to observation of actions made with tools in monkey ventral premotor cortex. *Journal of Cognitive Neuroscience*, 17: 212–26.

Fiorio, M., Weise, D., Önal-Hartmann, C., Zeller, D., Tinazzi, M., and Classen, J. (2011). Impairment of the rubber hand illusion in focal hand dystonia. *Brain*, 134(Pt 5): 1428–37.

First, M. B. (2005). Desire for amputation of a limb: paraphilia, psychosis, or a new type of identity disorder. *Psychological Medicine*, 35(6), 919–28.

Fodor, J. (1983). *The modularity of mind*. Cambridge, MA: MIT Press.

Folegatti, A., de Vignemont, F., Pavani, F., Rossetti, Y., and Farnè, A. (2009). Losing one's hand: visual-proprioceptive conflict affects touch perception. *PLoS One*, 4(9): e6920.

Folegatti, A., Farnè, A., Salemme, R., and de Vignemont, F. (2012). The Rubber Hand Illusion: two's a company, but three's a crowd. *Consciousness and Cognition*, 21(2): 799–812.

Fossataro, C., Gindri, P., Mezzanato, T., Pia, L., and Garbarini, F. (2016). Bodily ownership modulation in defensive responses: physiological evidence in brain-damaged patients with pathological embodiment of other's body parts. *Scientific Reports*, 6.

Fotopoulou, A., Jenkinson, P. M., Tsakiris, M., Haggard, P., Rudd, A., and Kopelman, M. D. (2011). Mirror-view reverses somatoparaphrenia: dissociation between first- and third-person perspectives on body ownership. *Neuropsychologia*, 49(14), 3946–55.

Fotopoulou, A., Pernigo, S., Maeda, R., Rudd, A., and Kopelman, M. A. (2010). Implicit awareness in anosognosia for hemiplegia: unconscious interference without conscious re-representation. *Brain*, 133(12), 3564–77.

Frassinetti, F., Ferri, F., Maini, M., Benassi, M. G., and Gallese, V. (2011). Bodily self: an implicit knowledge of what is explicitly unknown. *Experimental Brain Research*, 212(1), 153–60.

Frassinetti, F., Fiori, S., D'Angelo, V., Magnani, B., Guzzetta, A., Brizzolara, D., and Cioni, G. (2012). Body knowledge in brain-damaged children: a double-dissociation in self and other's body processing. *Neuropsychologia*, 50(1): 181–8.

Frassinetti, F., Maini, M., Romualdi, S., Galante, E., and Avanzi, S. (2008). Is it mine? Hemispheric asymmetries in corporeal self-recognition. *Journal of Cognitive Neuroscience*, 20(8): 1507–16.

Frances, A. and Gale, L. (1975). The proprioceptive body image in self-object differentiation—a case of congenital indifference to pain and head-banging. *Psychoanalytic Quarterly*, 44: 107–26.

Frege, G. (1956). The thought: a logical inquiry. *Mind*, 65, 259, 289–311.

Frijda, N. (2007). *The Laws of Emotion*. Mahwah, NJ: Lawrence Erlbaum.

Frith, C. (2005). The self in action: lessons from delusions of control. *Consciousness and Cognition*, 14(4), 752–70.

Fuchs, X., Riemer, M., Diers, M., Flor, H., and Trojan, J. (2016). Perceptual drifts of real and artificial limbs in the rubber hand illusion. *Scientific Reports*, 6:24362.

Fulkerson, M. (2014). *The first sense: a philosophical study of human touch.* Cambridge, MA: MIT Press.

Gallace, A. and Spence, S. (2010). Touch and the body: the role of the somatosensory cortex in tactile awareness. *Psyche*, 16(1), 31–67.

Gallagher, S. (1986). Body image and body schema: a conceptual clarification. *Journal of Mind and Behavior*, 7, 541–54.

Gallagher, S. (1995). Body schema and intentionality. In J. L. Bermúdez, A. Marcel, and N. Eilan (eds), *The body and the self.* Cambridge, MA: MIT Press.

Gallagher, S. (2000). Philosophical conceptions of the self: implications for cognitive science. *Trends in Cognitive Sciences*, 4(1), 14–21.

Gallagher, S. (2005). *How the body shapes the mind.* New York: Oxford University Press.

Gallagher, S. (2008). Are minimal representations still representations? *International Journal of Philosophical Studies*, 16(3), 351–69.

Gallagher, S. (2017). Enhancing the deflationary account of the sense of ownership. In F. de Vignemont and A. Alsmith (eds), *The subject's matter: Self-consciousness and the Body.* Cambridge, MA: MIT Press.

Gallagher, S. and Cole, J. (1995). Body schema and body image in a deafferented subject. *Journal of Mind and Behaviour*, 16, 369–90.

Gallagher, S. and Meltzoff, A. N. (1996). The earliest sense of self and others: Merleau-Ponty and recent developmental studies. *Philosophical Psychology*, 9, 211–33.

Gallese, V. (2001). The 'Shared Manifold' hypothesis: from mirror neurons to empathy. *Journal of Consciousness Studies*, 8(5–7): 33–50.

Gallese, V. (2007). Embodied simulation: from mirror neuron systems to interpersonal relations. *Novartis Foundation Symposium*, 278: 3–12.

Gallese, V. Fadiga, L., Fogassi, L., and Rizzolatti, G. (1996). Action recognition in the premotor cortex. *Brain*, 119, 593–609.

Gallese, V. and Sinigaglia, C. (2010). The bodily self as power for action. *Neuropsychologia*, 48(3), 746–55.

Gallese, V. and Sinigaglia, C. (2011). What is so special about embodied simulation? *Trends in Cognitive Sciences*, 15(11): 512–19.

Gandevia, S. C. and Phegan, C. M. (1999). Perceptual distortions of the human body image produced by local anaesthesia, pain and cutaneous stimulation. *Journal of Physiology*, 514, 609–16.

Gandola, M., Invernizzi, P., Sedda, A., Ferrè, E. R., Sterzi, R., Sberna, M., Paulesu, E., and Bottini, G. (2012). An anatomical account of somatoparaphrenia. *Cortex*, 48(9), 1165–78.

Garbarini, F., Fornia, L., Fossataro, C., Pia, L., Gindri, P., and Berti, A. (2014). Embodiment of others' hands elicits arousal responses similar to one's own hands. *Current Biology*, 24(16), R738–R739.

Garbarini, F., Fossataro, C., Berti, A., Gindri, P., Romano, D., Pia, L., della Gatta, F., Maravita, A., and Neppi-Modona, M. (2015). When your arm becomes mine: pathological embodiment of alien limbs using tools modulates own body representation. *Neuropsychologia*, 70: 402–13.

Garbarini, F., Pia, L., Fossataro, C., and Berti, A. (2017). From pathological embodiment to a model for body awareness. In F. de Vignemont and A. Alsmith (eds), *The subject's matter: Self-consciousness and the Body*. Cambridge, MA: MIT Press.

Garbarini, F., Rabuffetti, M., Piedimonte, A., Pia, L., Ferrarin, M., Frassinetti, F., Gindri, P., Cantagallo, A., Driver, J., and Berti, A. (2012). Moving a paralysed hand: bimanual coupling effect in patients with anosognosia for hemiplegia. *Brain*, 135, 5, 1486–97.

Gentile, G., Guterstam, A., Brozzoli, C., and Ehrsson, H. H. (2013). Disintegration of multisensory signals from the real hand reduces default limb self-attribution: an fMRI study. *The Journal of Neuroscience*, 33(33), 13350–66.

Georgieff, N. and Jeannerod, M. (1998). Beyond consciousness of external reality: a "Who" system for consciousness of action and self-consciousness. *Consciousness and Cognition*, 7(3): 465–77.

Gibson, J. J. (1979). *The ecological approach to visual perception*, Boston: Boston Mifflin.

Gillmeister, H., Catmur, C., Liepelt, R., Brass, M., and Heyes, C. (2008). Experience-based priming of body parts: a study of action imitation. *Brain Research*, 1217: 157–70.

Godfrey-Smith, P. (1994). A Modern History Theory of Functions. *Noûs*, 28, 344–62.

Gogol, N. (1835/1972). The nose. In *Diary of a madman and other stories*. London: Penguin Books.

Goldenberg, G. (1995). Imitating gestures and manipulating a mannikin—the representation of the human body in ideomotor apraxia. *Neuropsychologia*, 33(1): 63–72.

Goldenberg, G. (2009). Apraxia and the parietal lobes. *Neuropsychologia*, 47(6): 1449–59.

Goldman, A. I. (2006). *Simulating minds*. Oxford: Oxford University Press.

Goldman, A. I. (2012). A moderate approach to embodied cognitive science. In A. Alsmith and F. de Vignemont (eds), Special issue: The body represented/ Embodied representation. *Review of Philosophy and Psychology*, 3(1), 71–88.

Goldman, A. I. and de Vignemont, F. (2009). Is social cognition embodied? *Trends in Cognitive Sciences*, 13(4), 154–9.

Goldreich, D. and Kanics, I. M. (2003). Tactile acuity is enhanced in blindness. *Journal of Neuroscience*, 23(8), 3439–45.

González-Franco, M., Peck, T. C., Rodríguez-Fornells, A., and Slater, M. (2013). A threat to a virtual hand elicits motor cortex activation. *Experimental Brain Research*, 232(3), 875–87.

Goodale, M. A. and Westwood, D. A. (2004). An evolving view of duplex vision: separate but interacting cortical pathways for perception and action. *Current Opinion in Neurobiology*, 14(2), 203–11.

Gould, S. J. and Vrba, E. S. (1982). Exaptation—a missing term in the science of form. *Paleobiology*, 8(1), 4–15.

Gouzien, A. (2016). *Vers une théorie incarnée de la depression: étude expérimentale sur le lien entre l'état affectif, l'espace péripersonnel et les sensations corporelles*. Thèse de médecine. Université Paris XI.

Grafton, S. T., Fadiga, L., Arbib, M. A., and Rizzolatti, G. (1997). Premotor cortex activation during observation and naming of familiar tools. *Neuroimage*, 6(4): 231–6.

Grahek, N. (2001). *Feeling pain and being in pain*, Cambridge, MA: MIT Press.

Graydon, M. M., Linkenauger, S. A., Teachman, B. A., and Proffitt, D. R. (2012). Scared stiff: the influence of anxiety on the perception of action capabilities. *Cognition and Emotion*, 26(7): 1301–15.

Graziano, M. S. (2016). Ethological action maps: a paradigm shift for the motor cortex. *Trends in Cognitive Sciences*, 20(2), 121–32.

Graziano, M. S. A. and Gross, C. G. (1993). A bimodal map of space: somatosensory receptive fields in the macaque putamen with corresponding visual receptive fields. *Experimental Brain Research*, 97: 96–109.

Grezes, J. and Decety, J. (2001). Functional anatomy of execution, mental simulation, observation, and verb generation of actions: a meta-analysis. *Human Brain Mapping*, 12(1): 1–19.

Griffiths, T. L., Chater, N., Kemp, C., Perfors, A., and Tenenbaum, J. B. (2010). Probabilistic models of cognition: exploring representations and inductive biases. *Trends in Cognitive Sciences*, 14(8), 357–64.

Guardia, D., Conversy, L., Jardri, R., Lafargue, G., Thomas, P., Dodin, V., Cottencin, O., and Luyat, M. (2012). Imagining one's own and someone else's body actions: dissociation in anorexia nervosa. *PLoS One*, 7(8): e43241.

Guariglia, C. and Antonucci, G. (1992). Personal and extrapersonal space: a case of neglect dissociation. *Neuropsychologia*, 30(11), 1001–9.

Guariglia, C., Piccardi, L., Puglisi Allegra, M. C., and Traballesi, M. (2002). Is autotopoagnosia real? EC says yes. A case study. *Neuropsychologia*, 40, 1744–9.

Guillot, M. (2017). I me mine: on a confusion concerning the subjective character of experience. *Review of Philosophy and Psychology*, 8(1), 23–53.

Guillot, M. and Garcia-Carpintero, M. (eds) (forthcoming), *The sense of mineness*. Oxford: Oxford University Press.

Gurwitsch, A. (1985). *Marginal consciousness*. Athens OH: Ohio University Press.

Guterstam, A., Gentile, G., and Ehrsson, H. H. (2013). The invisible hand illusion: multisensory integration leads to the embodiment of a discrete volume of empty space. *Journal of Cognitive Neuroscience*, 25(7): 1078–99.

Guterstam, A., Petkova, V. I., and Ehrsson, H. H. (2011). The illusion of owning a third arm. *PLoS One*, 6(2): e17208.

Guterstam, A., Zeberg, H., Özçiftci, V. M., and Ehrsson, H. H. (2016). The magnetic touch illusion: A perceptual correlate of visuo-tactile integration in peripersonal space. *Cognition*, 155: 44–56.

Haggard, P. (2006). Just seeing you makes me feel better: interpersonal enhancement of touch. *Social Neuroscience*, 1(2): 104–10.

Haggard, P., Cheng, T., Beck, B., and Fardo, F. (2017). Spatial perception and the sense of touch. In F. de Vignemont and A. Alsmith (eds), *The subject's matter: self-consciousness and the body*. Cambridge, MA: MIT Press.

Hägni, K., Eng, K., Hepp-Reymond, M. C., Holper, L., Keisker, B., Siekierka, E., and Kiper, D. C. (2008). Observing virtual arms that you imagine are yours increases the galvanic skin response to an unexpected threat. *PloS One*, 3(8): e3082.

Halligan, P. W., Marshall, J. C., and Wade, D. T. (1995). Unilateral somatoparaphrenia after right hemisphere stroke: a case description. *Cortex*, 31(1), 173–82.

Head, H. (1920). Shell wound of head, right temporal region, sensory paresis of left hand and foot; mental and physical symptoms due to hole in skull; effect of closure with osteoplastic graft. *Proceedings of the Royal Society of Medicine*, 13 (Neurol Sect): 29–31.

Head, H. and Holmes, H. G. (1911). Sensory disturbances from cerebral lesions. *Brain*, 34, 102–254.

Hediger, H. (1950). *Wild animals in captivity*. London: Butterworths Scientific Publications.

Heed, T., Gründler, M., Rinkleib, J., Rudzik, F. H., Collins, T., Cooke, E., and O'Regan, J. K. (2011). Visual information and rubber hand embodiment differentially affect reach-to-grasp actions. *Acta psychologica*, 138(1), 263–71.

Helders, P. J. M. (1986). Early motor signs of blindness or very low vision in very young children. *Early Intervention*, 359–65.

Helm, B. (2002). Felt evaluations: a theory of pleasure and pain. *American Philosophical Quarterly*, 39(1): 13–30.

Helms Tillery, S., Stephen, I., Flanders, M., and Soechting, J. F. (1991). A coordinate system for the synthesis of visual and kinesthetic information. *Journal Neuroscience,* 11, 770–8.

Heyes, C. (2001). Causes and consequences of imitation. *Trends in Cognitive Sciences,* 5(6): 253–61.

Hilti, L. M., Hänggi, J., Vitacco, D. A., Kraemer, B., Palla, A., Luechinger, R., Jäncke, L., and Brugger, P. (2013). The desire for healthy limb amputation: structural brain correlates and clinical features of xenomelia. *Brain,* 136(Pt 1): 318–29.

Hlustik, P., Solodkin, A., Gullapalli, R. P., Noll, D. C., and Small, S. L. (2001). Somatotopy in human primary motor and somatosensory hand representations revisited. *Cerebral Cortex,* 11(4), 312–21.

Hochstetter, G. (2016). Attention in bodily awareness. *Synthese,* 193(12), 3819–42.

Hohwy, J. and Paton, B. (2010). Explaining away the body: experiences of supernaturally caused touch and touch on non-hand objects within the rubber hand illusion. *PLoS One,* 5(2):e9416.

Holly, W. T. (1986). The spatial coordinates of pain. *The Philosophical Quarterly,* 36(144), 343–56.

Holmes, N. P., Calvert, G. A., and Spence, C. (2007). Tool use changes multisensory interactions in seconds: evidence from the crossmodal congruency task. *Experimental Brain Research,* 183(4), 465–76.

Holmes, N. P., Crozier, G., and Spence, C. (2004). When mirrors lie: "visual capture" of arm position impairs reaching performance. *Cognitive, Affective and Behavioral Neuroscience,* 4(2): 193–200.

Holmes, N. P., Snijders, H. J., and Spence, C. (2006). Reaching with alien limbs: visual exposure to prosthetic hands in a mirror biases proprioception without accompanying illusions of ownership. *Perception and Psychophysics,* 68(4): 685–701.

Holmes, N. P. and Spence, C. (2005). Beyond the body schema: visual, prosthetic, and technological contributions to bodily perception and awareness. In G. Knoblich, I. M. Thornton, M. Grosjean, and M. Shiffrar (eds), *Human body perception from the inside out.* New York: Oxford University Press, 15–64.

Hurley, S. (1998). *Consciousness in action.* Cambridge, MA: Harvard University Press.

Hyvärinen, J. and Poranen, A. (1974). Function of the parietal associative area 7 as revealed from cellular discharges in alert monkeys. *Brain,* 97(4): 673–92.

Iacoboni, M., Molnar-Szakacs, I., Gallese, V., Buccino, G., Mazziotta, J. C., and Rizzolatti, G. (2005). Grasping the intentions of others with one's own mirror neuron system. *PLoS Biol,* 3(3): 529–35.

Invernizzi, P., Gandola, M., Romano, D., Zapparoli, L., Bottini, G., and Paulesu, E. (2013). What is mine? Behavioral and anatomical dissociations between somatoparaphrenia and anosognosia for hemiplegia. *Behavioral Neurology*, 261–2: 139–50.

Iriki, A., Tanaka, M., and Iwamura, Y. (1996). Coding of modified body schema during tool use by macaque postcentral neurones. *Neuroreport*, 7(14), 2325–30.

Ishida, H., Nakajima, K., Inase, M., and Murata, A. (2010). Shared mapping of own and others' bodies in visuotactile bimodal area of monkey parietal cortex. *Journal of Cognitive Neuroscience*, 22(1), 83–96.

Ishida, H., Suzuki, K., and Grandi, L. C. (2015). Predictive coding accounts of shared representations in parieto-insular networks. *Neuropsychologia*, 70, 442–54.

Jackson, P., Meltzoff, A., and Decety, J. (2005). How do we perceive the pain of others? A window into the neural processes involved in empathy. *Neuroimage*, 24: 771–9.

Jacob, F. (1977). Evolution and tinkering. *Science*, 196(4295), 1161–6.

Jacob, P. (2008). What do mirror neurons contribute to human social cognition? *Mind and Language*, 23(2), 190–223.

Jacob, P. and Jeannerod, M. (2003). *Ways of seeing, the scope and limits of visual cognition*. Oxford: Oxford University Press.

James, W. (1890). *The principles of psychology*, Vol 1. New York: Holt.

Jeannerod, M. (1997). *The cognitive neuroscience of action*. Oxford: Blackwell.

Jeannerod, M., Arbib, M. A., Rizzolatti, G., and Sakata, H. (1995). Grasping objects: the cortical mechanisms of visuomotor transformation. *Trends in Neuroscience*, 18(7): 314–20.

Jeannerod, M. and Pacherie, E. (2004). Agency, simulation and self-identification. *Mind and Language*, 19, 2: 113–46.

Jenkinson, P. M., Haggard, P., Ferreira, N. C., and Fotopoulou, A. (2013). Body ownership and attention in the mirror: insights from somatoparaphrenia and the rubber hand illusion. *Neuropsychologia*, 51(8): 1453–62.

Jones, B. (1972). Development of cutaneous and kinesthetic localization by blind and sighted children. *Developmental Psychology*, 6(2), pp. 349–52.

Jousmäki, V. and Hari, R. (1998). Parchment-skin illusion: sound-biased touch. *Current Biology*, 8(6), p. R190.

Kalckert, A. and Ehrsson, H. H. (2012). Moving a rubber hand that feels like your own: a dissociation of ownership and agency. *Frontiers in Human Neuroscience*, 6: 40.

Kalckert, A. and Ehrsson, H. H. (2014). The moving rubber hand illusion revisited: comparing movements and visuotactile stimulation to induce illusory ownership. *Consciousness and Cognition*, 26: 117–32.

Kammers, M. P., de Vignemont, F., Verhagen, L., and Dijkerman, H. C. (2009a). The rubber hand illusion in action. *Neuropsychologia*, 47, 204–11.

Kammers, M. P., Kootker, J. A., Hogendoorn, H., and Dijkerman, H. C. (2010). How many motoric body representations can we grasp? *Experimental Brain Research*, 202(1): 203–12.

Kammers, M., Mulder, J., de Vignemont, F., and Dijkerman, H. C. (2009b). The weight of representing the body: a dynamic approach to investigating multiple body representations in healthy individuals. *Experimental Brain Research*, 204(3), 333–42.

Kammers, M. P., Rose, K., and Haggard, P. (2011). Feeling numb: temperature, but not thermal pain, modulates feeling of body ownership. *Neuropsychologia*, 49(5): 1316–21.

Kammers, M. P., van der Ham, I. J., and Dijkerman, H. C. (2006). Dissociating body representations in healthy individuals: differential effects of a kinaesthetic illusion on perception and action. *Neuropsychologia*, 44, 2430–6.

Kandula, M., Hofman, D., and Dijkerman, H. C. (2015). Visuo-tactile interactions are dependent on the predictive value of the visual stimulus. *Neuropsychologia*, 70: 358–66.

Katz, D. (1925). *The world of touch*. Hillsdale, NJ: Psychology Press.

Kemmerer, D. (2014). Body ownership and beyond: connections between cognitive neuroscience and linguistic typology. *Consciousness and Cognition*, 26, 189–96.

Kennedy, D. P., Gläscher, J., Tyszka, J. M., and Adolphs, R. (2009). Personal space regulation by the human amygdala. *Nature Neuroscience*, 12(10), 1226–7.

Kennett, S., Spence, C., and Driver, J. (2002). Visuo-tactile links in covert exogenous spatial attention remap across changes in unseen hand posture. *Perception and Psychophysics*, 64(7), 1083–94.

Kennett, S., Taylor-Clarke, M., and Haggard, P. (2001). Noninformative vision improves the spatial resolution of touch in humans. *Current Biology*, 11(15), 1188–91.

Keysers, C., Wicker, B., Gazzola, V., Anton, J. L., Fogassi, L., and Gallese, V. (2004). A touching sight: SII/PV activation during the observation and experience of touch. *Neuron*, 42:335–46.

Khateb, A., Simon, S. R., Dieguez, S., Lazeyras, F., Momjian-Mayor, I., Blanke, O., Landis, T., Pegna, A. J., and Annoni, J. M. (2009). Seeing the phantom: a functional magnetic resonance imaging study of a supernumerary phantom limb. *Annals of Neurology*, 65(6): 698–705.

Kinsbourne, M. (1995). Awareness of one's own body: an attentional theory of its nature, development, and brain basis. In J. L. Burmúdez, A. J. Marcel, and N. Eilan (eds), *The body and the self*. Cambridge, MA: MIT Press, pp. 205–23.

Kinsbourne, M. (2002). The brain and body awareness. In T. F. Cash and T. Pruzinsky (eds), *Body image: a handbook of theory, research, and clinical practice* New York: Guildford Press, pp. 22–39.

Kinsbourne, M. and Lempert, H. (1980). Human figure representation by blind children. *The Journal of General Psychology*, 102, 33–7.

Klaver, M. and Dijkerman, H. C. (2016). Bodily experience in schizophrenia: factors underlying a disturbed sense of body ownership. *Frontiers in Human Neuroscience*, 10:305.

Klein, C. (2015a). *What the body commands*. Cambridge, MA: MIT Press.

Klein, C. (2015b). What pain asymbolia really shows. *Mind*, 124(494), 493–516.

Kóbor, I., Füredi, L., Kovács, G., Spence, C., and Vidnyánszky, Z. (2006). Back-to-front: improved tactile discrimination performance in the space you cannot see. *Neuroscience Letter*, 400(1–2), 163–7.

Kriegel, U. (2009). *Subjective consciousness: a self-representational theory*. Oxford: Oxford University Press.

Lackner, J. R. (1988). Some proprioceptive influences on the perceptual representation of body shape and orientation. *Brain*, 111, 281–97.

Lane, T. (2012). Toward an explanatory framework for mental ownership. *Phenomenology and the cognitive sciences*, 11(2), 251–86.

Lane, T., Yeh, S. L., Tseng, P., and Chang, A. Y. (2017). Timing disownership experiences in the rubber hand illusion. *Cognitive Research: Principles and Implications*, 2(1), 4.

Langdon, R. and Coltheart, M. (2000). The cognitive neuropsychology of delusions. *Mind and Language*, 15, 183–216.

Langdon, R., Connaughton, E., and Coltheart, M. (2014). The Fregoli delusion: a disorder of person identification and tracking. *Topics in Cognitive Science*, 6(4): 615–31.

Le Cornu Knight, F., Longo, M. R., and Bremner, A. J. (2014). Categorical perception of tactile distance. *Cognition*, 131(2), 254–62.

LeDoux, J. E. (2000). Emotion circuits in the brain. *Annual Review of Neuroscience*, 23:155–84.

Legrand, D. (2006). The bodily self: the sensori-motor roots of pre-reflective self-consciousness. *Phenomenology and the Cognitive Sciences*, 5(1), 89–118.

Leiguarda, R., Starkstein, S., Nogués, M., Berthier, M., and Arbelaiz, R. (1993). Paroxysmal alien hand syndrome. *Journal of Neurology, Neurosurgery, and Psychiatry*, 56(7): 788–92.

Lemon, R. (1988). The output map of the primate motor cortex. *Trends in Neuroscience*, 11(11), 501–6.

Lenggenhager, B., Tadi, T., Metzinger, T., and Blanke, O. (2007). Video ergo sum: manipulating bodily self-consciousness. *Science*, 317(5841), 109.6–109.9.

Levine, D. N., Calvanio, R., and Rinn, W. E. (1991). The pathogenesis of anosognosia for hemiplegia. *Neurology*, 41(11): 1770–81.

Levinson, S. C. (2006). Parts of the body in Yeli Dnye, the Papuan language of Rossel Island. *Language Sciences*, 28, 221–40.

Lewis, J. S., Kersten, P., McCabe, C. S., McPherson, K. M., and Blake, D. R. (2007). Body perception disturbance: a contribution to pain in complex regional pain syndrome (CRPS). *Pain*, 133(1), 111–19.

Lichstein L. and Sackett, G. P. (1971). Reactions by differentially raised rhesus monkeys to noxious stimulation. *Developmental Psychobiology*, 4(4): 339–52.

Limanowski, J. and Blankenburg, F. (2016). That's not quite me: limb ownership encoding in the brain. *Social Cognitive and Affective Neuroscience*, 11(7): 1130–40.

Lippman, C. W. (1952). Certain hallucinations peculiar to migraine. *Journal of Nervous and Mental Disease*, 116, 110–16.

Lloyd, D. M. (2007). Spatial limits on referred touch to an alien limb may reflect boundaries of visuo-tactile peripersonal space surrounding the hand. *Brain and Cognition*, 64(1): 104–9.

Locke, J. (1689/1997). *An essay concerning human understanding*. London: Penguin Classic.

Longhi, E., Senna, I., Bolognini, N., Bulf, H., Tagliabue, P., Macchi Cassia, V., and Turati, C. (2015). Discrimination of biomechanically possible and impossible hand movements at birth. *Child Development*, 86, 632–41.

Longo, M. R. (2017). Body representations and sense of self. In F. de Vignemont and A. Alsmith (eds), *The subject's matter: self-consciousness and the body*. Cambridge, MA: MIT Press.

Longo, M. R., Azanon, E., and Haggard, P. (2009b). More than skin deep: body representation beyond primary somatosensory cortex. *Neuropsychologia*, 48(3), pp. 655–68.

Longo, M. R. and Haggard, P. (2010). An implicit body representation underlying human position sense. *Proceedings of the National Academy of Sciences of the United States*, 107(26), 11727–32.

Longo, M. R. and Haggard, P. (2012). Implicit body representations and the conscious body image. *Acta Psychologica (Amst)*, 141(2): 164–8.

Longo, M. R., Schüür, F., Kammers, M. P., Tsakiris, M., and Haggard, P. (2008). What is embodiment? A psychometric approach. *Cognition*, 107: 978–98.

Longo, M. R., Schüür, F., Kammers, M. P., Tsakiris, M., and Haggard, P. (2009a). Self awareness and the body image. *Acta Psychologica (Amst)*, 132(2): 166–72.

Lotze, H. (1888). *Microcosmus: an essay concerning man and his relation to the world*. Princeton: Scribner and Welford.

Lourenco, S. F. and M. R. Longo (2009). The plasticity of near space: evidence for contraction. *Cognition*, 112(3): 451–6.

Lourenco, S. F., Longo, M. R., and Pathman, T. (2011). Near space and its relation to claustrophobic fear. *Cognition*, 119(3): 448–53.

Ma, K. and Hommel, B. (2013). The virtual-hand illusion: effects of impact and threat on perceived ownership and affective resonance. *Frontiers in Psychology*, 4, 604.

Ma, K. and Hommel, B. (2015a). The role of agency for perceived ownership in the virtual hand illusion. *Consciousness and Cognition*, 36: 277–88.

Ma, K. and Hommel, B. (2015b). Body-ownership for actively operated non-corporeal objects. *Consciousness and Cognition*, 36, 75–86.

Macpherson, F. (2011). Cross-modal experiences. *Proceedings of the Aristotelian Society*, 111(3), pp. 429–68.

Macpherson, F. (2012). Cognitive penetration of colour experience: rethinking the issue in light of an indirect mechanism. *Philosophy and Phenomenological Research*, 84(1), 24–62.

Majid, A., Enfield, N. J., and van Staden, M. (eds) (2006). Parts of the body: cross-linguistic categorisation [Special issue]. *Language Sciences*, 28(2–3).

Makin, T. R., Holmes, N. P., and Ehrsson, H. H. (2008). On the other hand: dummy hands and peripersonal space. *Behavioral and Brain Research*, 191(1), 1–10.

Malebranche, N. (1674/1980). *The search after truth*. Trans. T. M. Lennon. Columbus: Ohio State University Press.

Mamassian, P., Landy, M., and Maloney, L. T. (2002). Bayesian modelling of visual perception. In R. Rao, B. Olshausen, and M. Lewicki (eds), *Probabilistic models of the brain*, Cambridge, MA: MIT Press, 13–36.

Mancini, F., Longo, M. R., Iannetti, G. D., and Haggard, P. (2011). A supramodal representation of the body surface. *Neuropsychologia*, 49(5), 1194–201.

Maravita, A. (2008). Spatial disorders. In S. F. Cappa, J. Abutalebi, J. F. Demonet, P. C. Fletcher, and P. Garrard (eds), *Cognitive neurology: a clinical textbook*. New York: Oxford University Press, 89–118.

Maravita, A. and Iriki, A. (2004). Tools for the body (schema). *Trends in Cognitive Sciences*, 8(2), 79–86.

Maravita, A., Clarke, K., Husain, M., and Driver, J. (2002). Active tool use with the contralesional hand can reduce cross-modal extinction of touch on that hand. *Neurocase*, 8(6), 411–16.

Maravita, A., Spence, C., and Driver, J. (2003). Multisensory integration and the body schema: close to hand and within reach. *Current Biology*, 13(13): R531–9.

Marcel, A. (2003). The sense of agency: awareness and ownership of action. In J. Roessler and N. Eilan (eds), *Agency and self-awareness: issues in philosophy and psychology*. Oxford: Oxford University Press.

Marcel, A., Tegnér, R., and Nimmo-Smith, I. (2004). Anosognosia for plegia: specificity, extension, partiality and disunity of bodily unawareness. *Cortex*, 40(1), 19–40.

Marchetti, C. and Della Salla, S. (1998). Disentangling the alien and the anarchic hand. *Cognitive Neuropsychiatry*, 3, 191–207.

Margolis, J. (1966). Awareness of sensations and of the location of sensations. *Analysis*, 27, 29–32.

Marino, B. F. M., Stucchi, N., Nava, E., Haggard, P., and Maravita, A. (2010). Distorting the visual size of the hand affects hand pre-shaping during grasping. *Experimental Brain Research*, 202, pp. 499–505.

Marotta, A., Ferrè, E. R., and Haggard, P. (2015). Transforming the thermal grill effect by crossing the fingers. *Current Biology*, 25(8): 1069–73.

Marr, D. (1982). *Vision: a computational investigation into the human represen-tation and processing of visual information*. New York: W. H. Freeman and Company.

Marshall, J. C. and Halligan, P. W. (1988). Blindsight and insight in visuo-spatial neglect. *Nature*, 336, 766–7.

Marshall, J. C. and Halligan, P. W. (1994). The yin and the yang of visuo-spatial neglect: a case study. *Neuropsychologia*, 32(9): 1037–57.

Marshall, P. J. and Meltzoff, A. N. (2015). Body maps in the infant brain. *Trends in Cognitive Sciences*, 19(9), 499–505.

Martel, M., Cardinali, L., Roy, A. C., and Farnè, A. (2016). Tool-use: an open window into body representation and its plasticity. *Cognitive Neuropsychology*, 33(1–2), 82–101.

Martin, M. G. F. (1992). Sight and touch. In T. Crane (ed.), *The content of experience*. Cambridge: Cambridge University Press, 199–201.

Martin, M. G. F. (1993). Sense modalities and spatial properties. In N. Eilan, R. McCarty, and B. Brewer (eds), *Spatial representations*. Oxford: Oxford University Press, pp. 206–18.

Martin, M. G. F. (1995). Bodily awareness: a sense of ownership. In J. L. Bermúdez, T. Marcel, and N. Eilan (eds), *The body and the self*. Cambridge, MA: MIT Press.

Maselli, A., Kilteni, K., López-Moliner, J., and Slater, M. (2016). The sense of body ownership relaxes temporal constraints for multisensory integration. *Scientific Reports*, 6: 30628.

Massin, O. (2010). *L'objectivité du toucher*. PhD dissertation, Université Aix-Marseilles.

Matelli, M. and Luppino, G. (2001). Parietofrontal circuits for action and space perception in the macaque monkey. *Neuroimage*, 14(1), S27–S32.

Matthen, M. P. (2005). *Seeing, doing, and knowing: a philosophical theory of sense perception*. Oxford: Oxford University Press.

Matthen, M. P. (2006). On visual experience of objects. *Philosophical Studies*, 127, 195–220.

McDonnell, P. M., Scott, R. N., Dickison, J., Theriault, R. A., and Wood, B. (1989). Do artificial limbs become part of the user? New evidence. *Journal of Rehabilitation Research and Development*, 26(2): 17–24.

McDowell, J. (2011). Anscombe on bodily self-knowledge. *Essays on Anscombe's Intention*, edited by A. Ford, J. Hornsby, and F. Stoutland. Cambridge, MA: MIT Press.

McGlone, F., Wessberg, J., and Olausson, H. (2014). Discriminative and affective touch: sensing and feeling. *Neuron*, 82(4), 737–55.

McGonigle, D. J., Hanninen, R., Salenius, S., Hari, R., Frackowiak, R. S., and Frith, C. D. (2002). Whose arm is it anyway? An fMRI case study of supernumerary phantom limb. *Brain*, 125, 1265–74.

McIntosh, R. D., Brodie, E. E., Beschin, N., and Robertson, I. H. (2000). Improving the clinical diagnosis of personal neglect: a reformulated comb and razor test. *Cortex*, 36, 289–92.

Medina, J. and Coslett, H. B. (2016). What can errors tell us about body representations? *Cognitive Neuropsychology*, 33: 1–2, 5–25.

Meltzoff, A. N. (2007). 'Like me': a foundation for social cognition. *Developmental Science*, 10, 126–34.

Meltzoff, A. N. and Moore, M. K. (1995). Infants' understanding of people and things: from body imitation to folk psychology. In J. L. Bermúdez, A. Marcel, and N. Eilan (eds), *The body and self*. Cambridge, MA: MIT Press, pp. 43–69.

Melzack, R. (1990). Phantom limbs and the concept of a neuromatrix. *Trends in Neuroscience*, 13(3), 88–92.

Melzack, R. and Scott, T. H. (1957). The effects of early experience on the response to pain. *Journal of Comparative and Physiological Psychology*, 50(2): 155–61.

Melzack, R. and Wall, P. D. (1983). *The challenge of pain*. New York: Basic Books.

Melzack, R., Israel, R., Lacroix, R., and Schultz, G. (1997). Phantom limbs in people with congenital limb deficiency or amputation in early childhood. *Brain*, 120(9), pp. 1603–20.

Merabet, L. B. and Pascual-Leone, A. (2010). Neural reorganization following sensory loss: the opportunity of change. *Nature Review Neuroscience*, 11(1), pp. 44–52.

Merleau-Ponty, M. (1945). *Phénoménologie de la perception*. Paris: Gallimard.

Metzinger, T. (2006). Conscious volition and mental representation: toward a more fine-grained analysis. In N. Sebanz and W. Prinz (eds), *Disorders of volition*. Cambridge, MA: Bradford Books pp. 19–48.

Mezue, M. and Makin, T. (2017). Immutable body representations: lessons from phantoms in amputees. In F. de Vignemont and A. Alsmith (eds). *The subject's matter: self-consciousness and the body*. Cambridge, MA: MIT Press.

Miller, L. E., Longo, M. R., and Saygin, A. P. (2017). Visual illusion of tool use recalibrates tactile perception. *Cognition*, 162, 32–40.

Millikan, R. G. (1995). Pushmi-Pullyu Representations. *Philosophical Perspectives*, 9, 185–200.

Millikan, R. G. (1999). Wings, spoons, pills, and quills: a pluralist theory of function. *The Journal of Philosophy*, 96(4), 191–206.

Milner, A. D. and Goodale, M. A. (1995). *The visual brain in action*. Oxford: Oxford University Press.

Mitchell, S. W. (1871). Phantom limbs. *Lippincott's Magazine of Popular Literature and Science*, 8, 563–9.

Miyazaki, M., Hirashima, M., and Nozaki, D. (2010). The "cutaneous rabbit" hopping out of the body. *Journal of Neuroscience*, 305: 1856–60.

Mizumoto, M. and Ishikawa, M. (2005). Immunity to error through misidentification and the bodily illusion experiment. *Journal of Consciousness Studies*, 12(7), 3–19.

Moore, J. W. and Fletcher, P. C. (2012). Sense of agency in health and disease: a review of cue integration approaches. *Consciousness and Cognition*, 21(1): 59–68.

Moran, R. (2004). Anscombe on practical knowledge. *Royal Institute of Philosophy Supplement*, 55, 43–68.

Moro, V., Zampini, M., and Aglioti, S. M. (2004). Changes in spatial position of hands modify tactile extinction but not disownership of contralesional hand in two right brain-damaged patients. *Neurocase*, 10–16, 437–43.

Morrison, J. B. and Tversky, B. (2005). Bodies and their parts. *Memory and Cognition*, 33, 696–709.

Morse, J. M. and Mitcham, C. (1998). The experience of agonizing pain and signals of disembodiment. *Journal of Psychosomatic Research*, 44(6), 667–80.

Moseley, G. L. (2004). Why do people with complex regional pain syndrome take longer to recognize their affected hand? *Neurology*, 62(12): 2182–6.

Moseley, G. L. (2005). Distorted body image in complex regional pain syndrome. *Neurology*, 65, 773–1773.

Moseley, G. L., Gallace, A., and Iannetti, G. D. (2012b). Spatially defined modulation of skin temperature and hand ownership of both hands in patients with unilateral complex regional pain syndrome. *Brain*, 135(12), 3676–86.

Moseley, G. L., Gallace, A., and Spence, C. (2012a). Bodily illusions in health and disease: physiological and clinical perspectives and the concept of a cortical "body matrix". *Neuroscience and Biobehavioral Reviews*, 36, 34–46.

Moseley, G. L., Olthof, N., Venema, A., Don, S., Wijers, M., Gallace, A., and Spence, C. (2008). Psychologically induced cooling of a specific body part caused by the illusory ownership of an artificial counterpart. *Proceedings of the National Academy of Sciences*, 105: 13169–73.

Murray, C. D. (2004). An interpretative phenomenological analysis of the embodiment of artificial limbs. *Disability and Rehabilitation*, 26, 963–73.

Mylopoulos, M. (2015). Agentive awareness is not sensory awareness. *Philosophical Studies*, 172, 3, 761–80.

Nagasako, E. M., Oaklander, A. L., and Dworkin, R. H. (2003). Congenital insensitivity to pain: an update. *Pain*, 101(3): 213–19.

Nava, E., Steiger, T., and Röder, B. (2014). Both developmental and adult vision shape body representations. *Scientific Reports*, 4: 6622.

Neisser, U. and Becklen, R. (1975). Selective looking: attending to visually specified events. *Cognitive Psychology*, 7: 480–94.

Newport, R. and Gilpin, H. R. (2011). Multisensory disintegration and the disappearing hand trick. *Current Biology*, 21(19): R804–5.

Newport, R., Pearce, R., and Preston, C. (2010). Fake hands in action: embodiment and control of supernumerary limbs. *Experimental Brain Research*, 204: 385–95.

Nielsen, M. (1938). Gerstmann syndrome: Finger agnosia, agraphia, confusion of right and left and acalculia: comparison of this syndrome with disturbance of body scheme resulting from lesions of the right side of the brain. *Archives of Neurology and Psychiatry*, 39(3), 536–60.

Noë, A. (2004). *Action in perception*. Cambridge, MA: MIT Press.

Noë, A. (2005). Real presence. *Philosophical Topics*, 33, 235–64.

Noë, A. (2010). Vision without representation. In N. Gangopadhyay, M. Madary, and F. Spicer (eds), *Perception, action and consciousness*. Oxford: Oxford University Press.

Noordhof, P. (2001). In pain. *Analysis*, 61(2): 95–7.

O'Brien, L. (2003). On knowing one's own actions. In J. Roessler and N. Eilan (eds). *Agency and self-awareness*. Oxford: Clarendon Press, pp. 358–82.

O'Brien, L. (2007). *Self-knowing agents*. Oxford: Oxford University Press.

O'Callaghan, C. (2011). Perception and multimodality. In E. Margolis, R. Samuels, and S. Stich, *Oxford handbook of philosophy and cognitive science*. Oxford: Oxford University Press, pp. 92–117.

O'Regan, K. (2011). *Why red doesn't sound like a bell*. Oxford: Oxford University Press.

O'Regan, K. and Noë, A. (2001). A sensorimotor account of vision and visual consciousness. *Behavioral and Brain Sciences*, 24(5), 939–73.

O'Shaughnessy, B. (1980). *The will*. Vol. 1. Cambridge: Cambridge University Press.

O'Shaughnessy, B. (1989). The sense of touch. *Australasian Journal of Philosophy*, 67: 1, 37–58.

O'Shaughnessy, B. (1995). Proprioception and the body image. In J. L. Bermúdez, T. Marcel, and N. Eilan (eds), *The body and the self*. Cambridge, MA: MIT Press.

O'Shaughnessy, B. (2003). *Consciousness and the world*. Oxford: Oxford University Press.

Olkkonen, M., Hansen, T., and Gegenfurtner, K. R. (2008). Colour appearance of familiar objects: effects of object shape, texture and illumination changes. *Journal of Vision*, 8: 1–16.

Orbach, I., Stein, D., Palgi, Y., Asherov, J., Har-Even, D., and Elizur, A. (1996). Perception of physical pain in accident and suicide attempt patients: self-preservation vs self-destruction. *Journal of Psychiatric Research*, 30(4), 307–20.

Orwell, G. (1948). *1984*. London: Penguin.

Ovid (2004). *Metamorphoses*. London: Penguin Classic.

Pacherie, E. (2011). Self-agency. In S. Gallagher (ed.), *The Oxford handbook of the self*. Oxford: Oxford University Press, pp. 440–62.

Pagel, B., Heed, T., and Röder, B. (2009). Change of reference frame for tactile localization during child development. *Developmental Science*, 12(6), 929–37.

Paillard, J. (1980). Le corps situé et le corps identifié. Une approche psychophysiologique de la notion de schéma corporel. *Revue Médicale de la Suisse Romande*, 100, 129–41.

Paillard, J. (1999). Body schema and body image: a double dissociation in deafferented patients. In G. N. Gantchev, S. Mori, and J. Massion (eds), *Motor control, today and tomorrow*. Sofia: Academic Publishing House, pp. 197–214.

Paillard, J., Michel, F., and Stelmach, G. (1983). Localization without content. a tactile analogue of "blind sight". *Archives of Neurology*, 40, 548–51.

Paqueron, X., Leguen, M., Rosenthal, D., Coriat, P., Willer, J. C., and Danziger, N. (2003). The phenomenology of body image distortions induced by regional anesthesia. *Brain*, 126: 702–12.

Pavani, F. and Zampini M. (2007). The role of hand size in the fake-hand illusion paradigm. *Perception*, 36(10), 1547–54.

Peacocke, C. (1992). *A study of concepts*. Cambridge, MA: MIT Press.

Peacocke, C. (2003). Action: Awareness, ownership, and knowledge. In Johannes Roessler (ed.), *Agency and self-awareness: issues in philosophy and psychology*. Oxford: Clarendon Press.

Peacocke, P. (2012). Explaining De Se Phenomena. In S. Prosser and F. Recanati (eds), *Immunity to error through misidentification: new essays*. Cambridge: Cambridge University Press.

Peacocke, C. (2014). *The mirror of the world: subjects, consciousness, and self-consciousness*. Oxford: Oxford University Press.

Peacocke, C. (2015). Perception and the first person. In M. Matthen (ed.), *The Oxford handbook of the philosophy of perception*. Oxford: Oxford University Press.

Peacocke, C. (2017). Philosophical reflections on the first person. In F. de Vignemont and A. Alsmith (eds), *The subject's matter: self-consciousness and the body*. Cambridge, MA: MIT Press.

Peled, A., Pressman, A., Geva, A. B., and Modai, I. (2003). Somatosensory evoked potentials during a rubber-hand illusion in schizophrenia. *Schizophrenia Research*, 64: 157–63.

Penfield, W. and Rasmussen, T. (1950). *The cerebral cortex of man*. New York: MacMillan.

Perry, J. and Blackburn, S. (1986). Thought without representation. *Proceedings of the Aristotelian Society, Supplementary Volumes*, 60, 137–66.

Petkova, V. I. and Ehrsson, H. H. (2008). If I were you: perceptual illusion of body swapping. *PLoS One*, 3(12): e3832.

Petkova, V. I. and Ehrsson, H. H. (2009). When right feels left: referral of touch and ownership between the hands. *PLoS One*, 4: e6933.

Petkova, V. I., Zetterberg, H., and Ehrsson, H. H. (2012). Rubber hands feel touch, but not in blind individuals. *PLoS One*, 7(4): e35912.

Pia, L., Garbarini, F., Burin, D., Fossataro, C., and Berti, A. (2015). A predictive nature for tactile awareness? Insights from damaged and intact central-nervous-system functioning. *Frontiers in Human Neuroscience*, 19, 9: 287.

Pia, L., Garbarini, F., Fossataro, C., Fornia, L., and Berti, A. (2013). Pain and body awareness: evidence from brain-damaged patients with delusional body ownership. *Frontiers in Human Neuroscience*, 7: 298.

Pitron, V. and de Vignemont, F. (2017). Beyond differences between the body schema and the body image: insights from body hallucinations. *Consciousness and Cognition* 53, 115–21.

Poeck, K. and Orgass, B. (1964). On the development of the body image. Studies in normal children, blind children and child amputees. *Fortschr Neurol Psychiatr Grenzgeb*, 32, 538–55.

Poeck, K. and Orgass, B. (1971). The concept of the body schema: a critical review and some experimental results. *Cortex*, 7, 254–77.

Pons, T. P., Garraghty, P. E., Ommaya, A. K., Kaas, J. H., Taub, E., and Mishkin, M. (1991). Massive cortical reorganization after sensory deafferentation in adult macaques. *Science*, 252: 1857–60.

Posner, M. I., Snyder, C. R. R., and Davidson, B. J. (1980). Attention and the detection of signals. *Journal of Experimental Psychology: General*, 109: 160–74.

Pouget, A., Deneve, S., and Duhamel, J.-R. (2002). A computational perspective on the neural basis of multisensory spatial representations. *Nature Reviews Neuroscience*, 3, 741–7.

Povinelli, D. J., Reaux, J. E., and Frey, S. H. (2010). Chimpanzees' context-dependent tool use provides evidence for separable representations of hand

and tool even during active use within peripersonal space. *Neuropsychologia*, 48: 243–7.

Press, C., Taylor-Clarke, M., Kennett, S., and Haggard, P. (2004). Visual enhancement of touch in spatial body representation. *Experimental Brain Research*, 154(2): 238–45.

Preston, C. (2013). The role of distance from the body and distance from the real hand in ownership and disownership during the rubber hand illusion. *Acta Psychologica* (Amst), 142(2): 177–83.

Price, E. H. (2006). A critical review of congenital phantom limb cases and a developmental theory for the basis of body image. *Consciousness and Cognition*, 15(2), 310–22.

Pryor, J. (1999). Immunity to error through misidentification. *Philosophical Topics*, 26: 1, 271–304.

Pylyshyn, Z. (2001). Visual indexes, preconceptual objects, and situated vision. *Cognition*, 80, 127–58.

Pylyshyn, Z. (2003). *Seeing and visualizing: It's not what you think.* Cambridge, MA: MIT Press, A Bradford series book.

Raftopoulos, A. (2009). Reference, perception, and attention. *Philosophical Studies*, 144: 339–60.

Raftopoulos, A. and Muller, V. (2006). Nonconceptual demonstrative reference. *Philosophical and Phenomenological Research*, 72(2), 251–86.

Rahmanovic, A., Barnier, A. J., Cox, R. E., Langdon, R. A., and Coltheart, M. (2012). "That's not my arm": A hypnotic analogue of somatoparaphrenia. *Cognitive neuropsychiatry*, 17(1), 36–63.

Ramachandran, V. S. (1995). Anosognosia in parietal lobe syndrome. *Consciousness and Cognition*, 4, 22–51.

Ramachandran, V. S. (1998). Consciousness and body image: lessons from phantom limbs, Capgras syndrome and pain asymbolia. *Philosophical Transactions of the Royal Society of London. Series B: Biological Sciences*, 353(1377), 1851–9.

Ramachandran, V. S. and Blakeslee, S. (1998). *Phantoms in the brain.* London: Fourth Estate.

Ramachandran, V. S. and Hirstein, W. (1998). The perception of phantom limbs. *Brain*, 121, 1603–30.

Ramachandran, V. S. and McGeoch, P. (2007). Can vestibular caloric stimulation be used to treat apotemnophilia? *Medical Hypotheses*, 69(2): 250–2.

Ramachandran, V. S. and Rogers-Ramachandran, D. (1996). Synaesthesia in phantom limbs induced with mirrors. *Proceedings of the Royal Society of London*, 263: 377–86.

Reed, C. L. and Farah, M. J. (1995). The psychological reality of the body schema: a test with normal participants. *Journal of Experimental Psychology:Human Perception and Performance*, 21: 334–43.

Rey, G. (1977). Survival. In A. O. Rorty (ed), *The identities of persons*. Berkeley, CA: University of California Press, pp. 41–66.

Riemer, M., Bublatzkya, F., Trojanc, J., and Alpersaa, G. W. (2015). Defensive activation during the rubber hand illusion: Ownership versus proprioceptive drift. *Biological Psychology*, 109, 86–92.

Riemer, M., Kleinböhl, D., Hölzl, R., and Trojanc, J. (2013). Action and perception in the rubber hand illusion. *Experimental Brain Research*, 229(3), 383–93.

Rizzolatti, G. and Luppino, G. (2001). The cortical motor system. *Neuron*, 31(6), 889–901.

Rizzolatti, G., Fadiga, L., Fogassi, L., and Gallese, V. (1997). The space around us. *Science*, 277(5323), 190–1.

Rizzolatti, G., Fadiga, L., Gallese, V., and Fogassi, L. (1995). Premotor cortex and the recognition of motor actions. *Cognitive Brain Research*, 3, 131–41.

Rizzolatti, G., Scandolara, C., Matelli, M., and Gentilucci, M. (1981). Afferent properties of periarcuate neurons in macaque monkeys. II. Visual responses. *Behavioural Brain Research*, 2(2), 147–63.

Rochat, P. (1998). Self perception and action in infancy. *Experimental Brain Research*, 123, 102–9.

Rode, G., Charles, N., Perenin, M. T., Vighetto, A., Trillet, M., and Aimard, G. (1992). Partial remission of hemiplegia and somatoparaphrenia through vestibular stimulation in a case of unilateral neglect. *Cortex*, 28(2), 203–8.

Röder, B., Rösler, F., and Spence, C. (2004). Early vision impairs tactile perception in the blind. *Current Biology*, 14(2), 121–4.

Roessler, J. and Eilan, N. (eds) (2003). *Agency and self-awareness: issues in philosophy and psychology*. Oxford: Oxford University Press.

Rohde, M., Di Luca, M., and Ernst, M. O. (2011). The rubber hand illusion: feeling of ownership and proprioceptive drift do not go hand in hand. *PLoS One*, 6(6): e21659.

Rohde, M., Wold, A., Karnath, H. O., and Ernst, M. O. (2013). The human touch: skin temperature during the rubber hand illusion in manual and automated stroking procedures. *PLoS One*, 8(11): e80688.

Romano, D., Gandola, M., Bottini, G., and Maravita, A. (2014). Arousal responses to noxious stimuli in somatoparaphrenia and anosognosia: clues to body awareness. *Brain*, 137(Pt 4): 1213–23.

Romano, D., Sedda, A., Brugger, P., and Bottini, G. (2015). Body ownership: when feeling and knowing diverge. *Consciousness and Cognition*, 34, 140–8.

Rossetti, Y., Meckler, C., and Prablanc, C. (1994). Is there an optimal arm posture? Deterioration of finger localization precision and comfort sensation in extreme arm-joint postures. *Experimental Brain Research*, 99, 1301–36.

Rossetti, Y., Rode, G., and Boisson, D. (1995). Implicit processing of somaesthetic information: a dissociation between where and how? *Neuroreport*, 6, 506–10.

Rossetti, A., Romano, D., Bolognini, N., and Maravita, A. (2015). Dynamic expansion of alert responses to incoming painful stimuli following tool use. *Neuropsychologia*, 70, 486–94.

Russ, J. M., Shearin, E. N., Clarkin, J. F., Harrison, K., and Hull, J. W. (1993). Subtypes of self-injurious patients with borderline personality disorder. *The American Journal of Psychiatry*, 150, 1869–71.

Sacks, O. (1984). *A Leg to Stand On*. London: Picador.

Salomon, R., Noel, J. P., Łukowska, M., Faivre, N., Metzinger, T., Serino, A., and Blanke, O. (2017). Unconscious integration of multisensory bodily inputs in the peripersonal space shapes bodily self-consciousness. *Cognition*, 166, 174–83.

Salvato, G., Gandola, M., Veronelli, L., Agostoni, E. C., Sberna, M., Corbo, M., and Bottini, G. (2016). The spatial side of somatoparaphrenia: a case study. *Neurocase*, 22(2), 154–60.

Samad, M., Chung, A. J., and Shams L. (2015). Perception of body ownership is driven by Bayesian sensory inference. *PLoS One*, 10(2): e0117178.

Sambo, C. F. and Iannetti, G. D. (2013). Better safe than sorry? The safety margin surrounding the body is increased by anxiety. *The Journal of Neuroscience*, 33(35), 14225–30.

Scandola, M., Aglioti, S. M., Avesani, R., Bertagnoni, G., Marangoni, A., and Moro, V. (2017). Corporeal illusions in chronic spinal cord injuries. *Consciousness and Cognition*, 49, 278–90.

Scandola, M., Tidoni, E., Avesani, R., Brunelli, G., Aglioti, S. M., and Moro, V. (2014). Rubber hand illusion induced by touching the face ipsilaterally to a deprived hand: evidence for plastic "somatotopic" remapping in tetraplegics. *Frontiers in Human Neuroscience*, 8.

Schilder, P. (1935). *The image and appearance of the human body*. New York: International Universities Press.

Scholl, B. J. (2001). Objects and attention: the state of the art. *Cognition*, 80(1–2), 1–46.

Schwenkler, J. (2015). Understanding practical knowledge. *Philosophers' Imprint*, 15, 15, 1–32.

Schwoebel, J., Boronat, C. B., and Branch Coslett, H. (2002a). The man who executed "imagined" movements: evidence for dissociable components of the body schema. *Brain and Cognition*, 50, 1–16.

Schwoebel, J. and Coslett, H. B. (2005). Evidence for multiple, distinct representations of the human body. *Journal of Cognitive Neuroscience*, 17, 543–53.

Schwoebel, J., Coslett, H. B., Bradt, J., Friedman, R., and Dileo, C. (2002b). Pain and the body schema: effects of pain severity on mental representations of movement. *Neurology*, 59(5), 775–7.

Searle, J. (1983). *Intentionality: an essay in the philosophy of mind.* Cambridge: Cambridge University Press.

Sedda, A. and Bottini, G. (2014). Apotemnophilia, body integrity identity disorder or xenomelia? Psychiatric and neurologic etiologies face each other. *Neuropsychiatric Disease and Treatment*, 10: 1255–65.

Semenza, C. and Goodglass, H. (1985). Localization of body parts in brain injured subjects. *Neuropsychologia*, 23, 161–75.

Senna, I., Maravita, A., Bolognini, N., and Parise, C. V. (2014). The Marble-Hand Illusion. *PLoS One*, 9(3): e91688.

Serino, A., Annella, L., and Avenanti, A. (2009). Motor properties of peripersonal space in humans. *PLoS One*, 4(8): e6582.

Serino, A., Bassolino, M., Farnè, A., and Làdavas, E. (2007). Extended multisensory space in blind cane users. *Psychological Science*, 18(7), 642–8.

Serrahima, C. (forthcoming). The bounded body: on the sense of bodily ownership and the experience of space. In M. Guillot and M. Garcia-Carpintero (eds), *The sense of mineness*. Oxford: Oxford University Press.

Seth, A. (2013). Interoceptive inference, emotion, and the embodied self, *Trends in Cognitive Science*, 17(11): 565–73.

Sforza, A., Bufalari, I., Haggard, P., and Aglioti, S. M. (2010). My face in yours: visuo-tactile facial stimulation influences sense of identity. *Social Neuroscience*, 5(2): 148–62.

Shams, L. and Kim, R. (2010). Crossmodal influences on visual perception. *Physics of Life Reviews*, 7: 269–84.

Shapiro, L. A. (2010). *Embodied cognition.* New York: Routledge.

Shapiro, M. F., Fink, M., and Bender, M. B. (1952). Exosomesthesia or displacement of cutaneous sensation into extrapersonal space. *AMA Archives of Neurology and Psychiatry*, 684: 481–90.

Sherrington, C. S. (1906). *The integrative action of the nervous system.* New Haven: Yale University Press.

Shoemaker, S. (1968). Self-reference and self-awareness. *The Journal of Philosophy*, 65, 555–67.

Siegel, S. (2009). The visual experience of causation. *Philosophical Quarterly*, 59(236), 519–40.

Siegel, S. (2010). *The contents of visual experience.* New York: Oxford University Press.

Sierra, M. (2009). *Depersonalization: a new look at a neglected syndrome.* Cambridge: Cambridge University Press.

Sierra, M. and Berrios, G. E. (2001). The phenomenological stability of depersonalization: comparing the old with the new. *The Journal of Nervous and Mental Disease*, 189(9): 629–36.

Siewert, C. (2005). Attention and sensorimotor intentionality. In D. Woodruff Smith and A. Thomasson (eds), *Phenomenology and philosophy of mind*. Oxford: Clarendon Press.

Simmons, A. (2008). Guarding the body: a Cartesian phenomenology of perception. In *Contemporary perspectives on early modern philosophy: essays in honor of Vere Chappell*, eds Paul Hoffman and Gideon Yaffe. Peterborough, ON: Broadview Press: pp. 81–113.

Singer, T., Seymour, B., O'Doherty, J., Kaube, H., Dolan, R., and Frith, C. (2004). Empathy for pain involves the affective but not sensory components of pain. *Science*, 303: 1157–62.

Sirigu, A., Grafman, J., Bressler, K., and Sunderland, T. (1991). Multiple representations contribute to body knowledge processing. Evidence from a case of autotopagnosia. *Brain*, 114, 629–42.

Slater, M., Perez-Marcos, D., Ehrsson, H. H., and Sanchez-Vives, M. V. (2008). Towards a digital body: the virtual arm illusion. *Frontiers in Human Neuroscience*, 2: 6.

Slimani, H., Danti, S., Ptito, M., and Kupers, R. (2014). Pain perception is increased in congenital but not late onset blindness. *PLoS One*, 9(9): e107281.

Smeets, J. B., van den Dobbelsteen, J. J., de Grave, D. D., van Beers, R. J., and Brenner, E. (2006). Sensory integration does not lead to sensory calibration. *Proceedings of the National Academy of Sciences of the United States*, 103(49), 18781–6.

Smith, A. D. (2002). *The problem of perception*. Cambridge, MA: Harvard University Press.

Smith, A. J. T. (2009). Acting on (bodily) experience. *Psyche*, 15(1).

Soliman, T. M., Ferguson, R., Dexheimer, M. S., and Glenberg, A. M. (2015). Consequences of joint action: entanglement with your partner. *Journal of Experimental Psychology: General*, 144(4), 873–88.

Sorene, E. D., Heras-Palou, C., and Burke, F. D. (2006). Self-amputation of a healthy hand: a case of body integrity identity disorder. *The Journal of Hand Surgery British and European*, 31(6): 593–5.

Spence, C. and Bayne, T. (2015). Is consciousness multisensory? In D. Stokes, S. Biggs, and M. Matthen (eds), *Perception and its modalities*. New York: Oxford University Press, pp. 95–132.

Spence, C., Pavani, F., and Driver, J. (2004). Spatial constraints on visual-tactile cross-modal distractor congruency effects. *Cognitive, Affective, and Behavioral Neuroscience*, 4(2), 148–69.

Spicker, S. F. T. (1975). The lived body as catalytic agent: reaction at the interface of medicine and philosophy. In H. R. Engelhardt and S. F. Spicker (eds), *Evaluation and explanation in the biomedical sciences*. Dordrecht: Reidel Publishing Co.

Sposito, A., Bolognini, N., Vallar, G., and Maravita, A. (2012). Extension of perceived arm length following tool-use: clues to plasticity of body metrics. *Neuropsychologia*, 50(9): 2187–94.

Staub, F., Bogousslavsky, J., Maeder, P., Maeder-Ingvar, M., Fornari, E., Ghika, J., Vingerhoets, F., and Assal, G. (2006). Intentional motor phantom limb syndrome. *Neurology*, 67(12), 2140–6.

Stein, B. E. and Meredith, M. A. (1993). *The merging of the senses*. Cambridge, MA: MIT Press.

Sterr, A., Green, L., and Elbert, T. (2003). Blind Braille readers mislocate tactile stimuli. *Biological Psychology*, 63(2), pp. 117–27.

Stratton, G. M. (1899). The spatial harmony of touch and sight. *Mind*, 8, 492–505.

Synofzik, M., Vosgerau, G., and Newen, A. (2008). I move, therefore I am: a new theoretical framework to investigate agency and ownership. *Consciousness and Cognition*, 17(2), 411–24.

Takaiwa, A., Yoshimura, H., Abe, H., and Terai, S. (2003). Radical "visual capture" observed in a patient with severe visual agnosia. *Behavioural Neurology*, 14(1–2): 47–53.

Tastevin, J. (1937). En partant de l'expérience d'Aristotle. *L'Encéphale*, 1, 140–58.

Taylor-Clarke, M., Jacobsen, P., and Haggard, P. (2004). Keeping the world a constant size: object constancy in human touch. *Nature Neuroscience*, 7(3), 219–20.

Thelen, E. and Smith, L. (1994). *A Dynamic Systems Approach to the Development of Cognition and Action*. Cambridge, MA: MIT Press.

Thomas, R., Press, C., and Haggard, P. (2006). Shared representations in body perception. *Acta Psychologica* (Amst), 121(3): 317–30.

Thompson, E. (2005). Sensorimotor subjectivity and the enactive approach to experience. *Phenomenology and the Cognitive Sciences*, 4(4), 407–27.

Tieri, G., Tidoni, E., Pavone, E. F., and Aglioti, S. M. (2015a). Body visual discontinuity affects feeling of ownership and skin conductance responses. *Scientific Reports*, 5: 17139.

Tieri, G., Tidoni, E., Pavone, E. F., and Aglioti, S. M. (2015b). Mere observation of body discontinuity affects perceived ownership and vicarious agency over a virtual hand. *Experimental Brain Research*, 233: 1247–59.

Tipper, S. P., Lloyd, D., Shorland, B., Dancer, C., Howard, L. A., and McGlone, F. (1998). Vision influences tactile perception without proprioceptive orienting. *Neuroreport*, 9, 1741–4.

Titchener, E. (1908). *Lectures on the elementary psychology of feeling and attention*. New York: Macmillan.

Todd, J. (1955). The syndrome of Alice in Wonderland. *Canadian Medical Association Journal*, 73(9): 701.

Torrance, F. (2011). Experience: I feel other people's pain. *Guardian*. 19 March 2011.

Tranel, D. and Damasio, A. R. (1985). Knowledge without awareness: an autonomic index of facial recognition by prosopagnosics. *Science*, 228, 1453–5.

Tranel, D. and Damasio, H. (1989). Intact electrodermal skin conductance responses after bilateral amygdala damage. *Neuropsychologia*, 27(4), 381–90.

Travieso, D., Pilar, M. A., and Gomila, A. (2007). Haptic perception is a dynamic system of cutaneous, proprioceptive, and motor components. *Behavioral and Brain Sciences*, 30, 222–3.

Treisman, A. (1998). Feature binding, attention and object perception. *Philosophical transactions of the Royal Society of London. Series B*, 353(1373), 1295–306.

Treisman, A. (1999). Feature binding, attention and object perception. In G. Humphreys, J. Duncan, and A. Treisman (eds), *Attention, Space, and Action*. Oxford: Oxford University Press.

Tsakiris, M. (2008). Looking for myself: current multisensory input alters self-face recognition. *PLoS One*, 3(12): e4040.

Tsakiris, M. (2010). My body in the brain: a neurocognitive model of body-ownership. *Neuropsychologia*, 48: 703–12.

Tsakiris, M. (2017). The material me. In F. de Vignemont and A. Alsmith (eds), *The subject's matter: self-consciousness and the body*. Cambridge, MA: MIT Press.

Tsakiris, M., Carpenter, L., James, D., and Fotopoulou, A. (2010). Hands only illusion: multisensory integration elicits sense of ownership for body parts but not for non-corporeal objects. *Experimental Brain Research*, 204(3): 343–52.

Tsakiris, M., Prabhu, G., and Haggard, P. (2006). Having a body versus moving your body: how agency structures body-ownership. *Consciousness and Cognition*, 15(2): 423–32.

Tucker, M. and Ellis, R. (1998). On the relations between seen objects and components of potential actions. *Journal of Experimental Psychology: Human Perception and Performance*, 24(3), 830–46.

Türker, K. S., Yeo, P. L., and Gandevia, S. C. (2005). Perceptual distortion of face deletion by local anaesthesia of the human lips and teeth. *Experimental Brain Research*, 165(1), 37–43.

Turvey, M. and Carello, C. (1995). Some dynamical themes in perception and action. In R. Port and T. van Gelder (eds), *Mind as Motion*. Cambridge, MA: MIT Press.

Tye, M. (2002). On the location of a pain. *Analysis*, 62(2): 150–3.

Vallar, G., Bottini, G., Sterzi, R., Passerini, D., and Rusconi, M. L. (1991). Hemianesthesia, sensory neglect, and defective access to conscious experience. *Neurology*, 41(5): 650–2.

Vallar, G. and Ronchi, R. (2009). Somatoparaphrenia: a body delusion. A review of the neuropsychological literature. *Experimental Brain Research*, 192(3), 533–51.

van Beers, R. J., Sittig, A. C., and Denier van der Gon, J. J. (1998). The precision of proprioceptive position sense. *Experimental Brain Research*, 122, pp. 367–77.

van Beers, R. J., Sittig, A. C., and Denier van der Gon, J. J. (1999). Integration of proprioceptive and visual position information: an experimentally supported model. *Journal of Neurophysiology*, 81(3), 1355-64.

van Beers, R. J., Wolpert, D. M., and Haggard, P. (2002). When feeling is more important than seeing in sensorimotor adaptation. *Current Biology*, 12(10), pp. 834-7.

van Gelder, T. (1995). What might cognition be, if not computation? *Journal of Philosophy*, XCII, 345-81.

Venneri, A., Pentore, R., Cobelli, M., Nichelli, P., and Shanks, M. F. (2012). Translocation of the embodied self without visuospatial neglect. *Neuropsychologia*, 50(5): 973-8.

Vesey, G. N. A. (1961). The location of bodily sensations. *Mind*, LXX277: 25-35.

de Vignemont, F. (2007). Habeas corpus: the sense of ownership of one's own body. *Mind and Language*, 22(4), 427-49.

de Vignemont, F. (2010). Body schema and body image: pros and cons. *Neuropsychologia*, 48(3), 669-80.

de Vignemont, F. (2011). A mosquito bite against the enactive view to bodily experiences. *Journal of Philosophy*, CVIII, 4, 188-204.

de Vignemont, F. (2012). Bodily immunity to error. In.F. Recanati and S. Prosser (eds), *Immunity to error through misidentification*. Cambridge: Cambridge University Press, 224-46.

de Vignemont, F. (2013). The mark of bodily ownership. *Analysis*, 73(4): 643-51.

de Vignemont, F. (2014a). A multimodal conception of bodily awareness. *Mind*, 123(492): 989-1020.

de Vignemont, F. (2014b). Multimodal unity and multimodal binding. In D. Bennett and C. Hill (eds), *Sensory integration and the unity of consciousness*. Cambridge, MA: MIT Press, 125-50.

de Vignemont, F. (2014c). Shared body representations and the "Whose" system. *Neuropsychologia*, 55, 128-36.

de Vignemont, F. (2016). Bodily experiences and bodily affordances. In Y. Coello and M. H. Fisher (eds), *Foundations of embodied cognition*. Hillsdale, NJ: Psychology Press.

de Vignemont, F. (2017a). The extended body hypothesis. In A. Newen, L. de Bruin, and S. Gallagher (eds), *Oxford handbook of 4E cognition*. Oxford: Oxford University Press.

de Vignemont, F. (2017b). Pain and touch. *The Monist*, 100, 4, 465-477.

de Vignemont, F. (2017c). 'Agency and bodily ownership'. In F. de Vignemont and A. Alsmith (eds), The Subject's Matter: *Self-consciousness and the Body*. Cambridge, MA: MIT Press.

de Vignemont, F., Erhsson, H., and Haggard, P. (2005a). Bodily illusions modulate tactile perception. *Current Biology*, 15(14): 1286-90.

de Vignemont, F. and Farnè, A. (2010). Widening the body to rubber hands and tools: What's the difference? *Revue de Neuropsychologie, Neurosciences Cognitives et Cliniques*, 2(3): 1–9.

de Vignemont, F. and Iannetti, G. (2015). How many peripersonal spaces? *Neuropsychologia*, 70: 327–34.

de Vignemont, F. and Jacob, P. (2012). What it's like to feel another's pain. *Philosophy of Science*, 79, 2, 295–316.

de Vignemont, F. and Jacob, P. (2016). Beyond empathy for pain. *Philosophy of Science*, 83: 3, 434–45.

de Vignemont, F., Majid, A., Jolla, C., and Haggard, P. (2009). Segmenting the body into parts: evidence from biases in tactile perception. *Quaterly Journal of Experimental Psychology*, 62(3): 500–12.

de Vignemont, F. and Massin, O. (2015). Touch. In M. Matthen (ed.), *Oxford handbook of philosophy of perception*. Oxford: Oxford University Press.

de Vignemont, F., Tsakiris, M., and Haggard, P. (2005b). Body mereology. In G. Knoblich, I. M. Thornton, M. Grosjean, and M. Shiffrar (eds), *Human body perception from the inside out*. New York: Oxford University Press, 147–70.

Von Békésy, G. (1959). Similarities between hearing and skin sensations. *Psychol Rev*, 661: 1–22.

Von Békésy, G. (1967). *Sensory inhibition*. Princeton: Princeton University Press.

Vuilleumier, P. (2004). Anosognosia: the neurology of beliefs and uncertainties. *Cortex*, 40(1), 9–17.

Vuilleumier, P., Reverdin, A., and Landis, T. (1997). Four legs. Illusory reduplication of the lower limbs after bilateral parietal lobe damage. *Archives in Neurology*, 54(12): 1543–7.

Ward, J. and Banissy, M. J. (2015). Explaining mirror-touch synesthesia. *Cognitive Neuroscience*, 6(2-3): 118–33.

Weber, E. H. (1826/1978). *The sense of touch* (*De Tactu*, H. E. Ross, trans.; *Der Tastsinn*, D. J. Murray, trans.). New York: Academic Press.

Weinstein, E. A., Kahn, R. L., Malitz, S., and Rozanski, J. (1954). Delusional reduplication of parts of the body. *Brain*, 77(1): 45–60.

Welch, R. B. (1999). Meaning, attention, and the "Unity Assumption" in the intersensory bias of spatial and temporal perceptions. *Advances in Psychology*, 129, pp. 371–87.

Welch, R. B. and Warren, D. H. (1980). Immediate perceptual response to intersensory discrepancy. *Psychological Bulletin*, 88, 638–67.

Wheaton, K. J., Thompson, J. C., Syngeniotis, A., Abbott, D. F., and Puce, A. (2004). Viewing the motion of human body parts activates different regions of premotor, temporal, and parietal cortex. *Neuroimage*, 22(1): 277–88.

Wiggins, D. (2001). *Sameness and substance renewed*. Cambridge: Cambridge University Press.

Wittgenstein, L. (1958). *The Blue and Brown Books*. Oxford: Blackwell.

Wolpert, D. M., Ghahramani, Z., and Flanagan, J. R. (2001). Perspectives and problems in motor learning. *Trends in Cognitive Science*, 5, 487–94.

Wong, H. Y. (2009). On the necessity of bodily awareness for bodily action. *Psyche*, 15(1).

Wong, H. Y. (2014). On the multimodality of body perception in action. *Journal of Consciousness Studies*, 21, 130–9.

Wong, H. Y. (2015). On the significance of bodily awareness for action. *Philosophical Quarterly*, 65(261): 790–812.

Wu, W. (forthcoming). Mineness and introspective data. In M. Guillot and M. Garcia-Carpintero (eds), *The sense of mineness*. Oxford: Oxford University Press.

Wundt, W. M. (1897). *Outlines of psychology*. Leipzig: G. E. Stechert.

Yamamoto, S. and Kitazawa, S. (2001a). Reversal of subjective temporal order due to arm crossing. *Nature Neuroscience*, 4(7), 759–65.

Yamamoto, S. and Kitazawa, S. (2001b). Sensation at the tips of invisible tools. *Nature Neuroscience*, 4(10): 979–80.

Yamamoto, S., Moizumi, S., and Kitazawa, S. (2005). Referral of tactile sensation to the tips of l-shaped sticks. *Journal of Neurophysiology*, 93, 2856–63.

Yoshimura, A., Matsugi, A., Esaki, Y., Nakagaki, K., and Hiraoka, K. (2010). Blind humans rely on muscle sense more than normally sighted humans for guiding goal-directed movement. *Neuroscience Letter*, 471(3).

Zeller, D., Gross, C., Bartsch, A., Johansen-Berg, H., and Classen, J. (2011). Ventral premotor cortex may be required for dynamic changes in the feeling of limb ownership: a lesion study. *Journal of Neuroscience*, 31(13): 4852–7.

Zhang, J. and Hommel, B. (2016). Body ownership and response to threat. *Psychological Research*, 80(6): 1020–9.

Zhu, L., Pei, G., Gu, L., and Hong, J. (2002). Psychological consequences derived during process of human hand allograft. *Chinese Medical Journal*, 115(11), 1660–3.

Zingerle, H. (1913). Uber Störungen der Wahrnehmung des eigenen Körpers bei organischen Gehirnerkrankungen. *Monatsschrift für Psychiatrie und Neurologie*, 34, 13–36.

Index